£3 00

# PHYSICS OF SOLID STATE DEVICES
T.H.Beeforth and H.J.Goldsmid

**Applied physics series**

Series editor H.J.Goldsmid

# PHYSICS OF SOLID STATE DEVICES
## T.H.Beeforth and H.J.Goldsmid

**Pion Limited, 207 Brondesbury Park, London NW2**

Library edition      SBN 85086 013 X
Student edition      SBN 85086 014 8

Set on IBM 72 Composers by Pion Limited, London.
Printed in Great Britain by J.W.Arrowsmith Limited, Bristol.

# Preface

The development of solid state devices during the past twenty or so years provides an outstanding example of the way that physics research can transform an industry. The revolution in electronic engineering has not only consisted of the replacement of most kinds of vacuum valve by semiconducting diodes and transistors: there has been the appearance of many new components made from advanced dielectric and magnetic materials, and there has been a rapid growth of what is now broadly called photoelectronics, which includes such diverse topics as techniques for the transmission of information by light waves, solid state display systems, and solar energy conversion. Nor should it be forgotten that the development of semiconductor junction devices has been accompanied by that of devices using homogeneous semiconductors. In addition, a whole new range of electronic applications based on superconductors and their fascinating behaviour has opened up because of the ready availability of liquid helium.

Not surprisingly, there is a widespread thirst for knowledge of the principles by which all these devices operate, and it is in an attempt to satisfy this thirst that we have written this book. There are, of course, many other books that deal with individual devices and a few that cover the whole field, but the latter are, on the whole, both long and expensive, since they go right back to the fundamental physics of the solid state. We have been able to compile a much shorter and cheaper book by omitting the basic solid state physics; after all, there are several excellent fundamental texts and it would be presumptuous of us to try to improve on them. Thus, we assume that our readers will be armed with at least a qualitative knowledge of the general outlines of solid state theory. We expect that our book will prove suitable for advanced undergraduates and for qualified scientists and engineers who require an introduction to solid-state device physics.

T.H.Beeforth
H.J.Goldsmid

# Contents

The design of the dust jacket is based on a photograph kindly supplied by
Marconi–Elliott Microelectronics Limited.

# Transport effects in semiconductors

## 1.1 Thermistors

The simplest semiconductor device is undoubtedly the thermistor. This is a type of temperature-sensitive resistor, which takes advantage of the fact that the proportional charge of electrical conductivity with temperature is far larger for a semiconductor than for a metal. It is also easier to obtain a conveniently high value of the electrical resistance using a semiconducting sample.

The electrical conductivity $\sigma$ of a pure metal, at ordinary temperatures, varies inversely with the absolute temperature $T$. This is because the conduction electrons are scattered by the lattice vibrations, which become more intense as the temperature rises. The electron concentration is independent of temperature, but the relaxation time at the Fermi energy varies as $1/T$. At temperatures that are much less than the Debye temperature $\theta_D$, the electrical conductivity certainly becomes more strongly temperature-dependent (simple theory indicates that $\sigma \propto T^{-5}$) and, of course, if the metal is a superconductor below some critical temperature $T_c$, $\sigma$ can become infinite. However, $T_c$ is never greater than about $20°K$, and the strong temperature dependence of the conductivity in the normal state is also seen only at quite low temperatures.

Ordinary resistors are made from alloys rather than pure metals, since alloy scattering considerably increases the resistivity. It also makes the resistivity much less dependent on temperature, which is an advantage in other applications but which makes metallic alloys even less suitable than metals for temperature-sensitive resistors.

Some semiconducting materials display the same weak temperature dependence of the conductivity as metals. These are materials in which the electrons or holes originate from impurity states, the impurities being fully ionised at the temperature of observation. Since the concentration of charge carriers does not then depend on temperature, the electrical conductivity varies only with the carrier relaxation time, which falls as the temperature rises if lattice scattering is dominant. At very low temperatures, the carrier concentration normally depends on the temperature, provided that the impurity levels are separated in energy from the conduction or valence band. The carrier concentration then becomes zero at $0°K$, but it would remain finite if there were a large number of impurities in the semiconductor. This is because, in a very impure semiconductor, the impurity levels broaden into a band which overlaps one or other of the main bands.

The carrier concentration also varies with temperature when the impurity concentration is low enough, or the temperature high enough, for the semiconductor to become intrinsic. It is, in fact, the exponential variation of electrical conductivity, or more strictly carrier concentration, with temperature, that is the most characteristic property of a semi-conductor.

In a thermistor material one generally aims at a negative coefficient of the resistivity of up to about 5% degK$^{-1}$, compared with a positive coefficient of about $\frac{1}{3}$% degK$^{-1}$ (near room temperature) for a pure metal. Such a rapid change of conductivity with temperature could be achieved using an intrinsic semiconductor for which the electron or hole concentration would be given by

$$n_i = 2\left(\frac{2\pi mkT}{h^2}\right)^{3/2}\left(\frac{m_n^* m_p^*}{m^2}\right)^{3/4}\exp\left(-\frac{E_g}{2kT}\right), \tag{1.1}$$

where $m$ is the mass of a free electron, $m_n^*$ the density-of-states effective mass for electrons, $m_p^*$ the density-of-states mass for holes, and $E_g$ the energy gap. $k$ and $h$ are Boltzmann's and Planck's constants respectively. The electrical conductivity is

$$\sigma = n_i e(\mu_n + \mu_p), \tag{1.2}$$

where $e$ is the electronic charge and $\mu_n$ and $\mu_p$ are the mobilities of electrons and holes. The exponential term in Equation (1.1) is expected to be the dominant factor in the variation of the conductivity with temperature. Actually, for acoustic-mode lattice scattering in a simple band, the theory indicates that the mobility varies as $T^{-3/2}$; then, combining Equations (1.1) and (1.2) we see that the electrical conductivity should vary strictly as $\exp(-E_g/2kT)$. Thus for an intrinsic semiconductor

$$\left(\frac{1}{\sigma}\right)\frac{d\sigma}{dT} = \frac{E_g}{2kT^2}. \tag{1.3}$$

It is difficult to obtain intrinsic conductivity at a temperature that is less than that for which $E_g/kT \approx 24$ (for example, very pure germanium with less than about $10^{19}$ donors m$^{-3}$ or acceptors m$^{-3}$ is needed to achieve this condition and such purity cannot be reached for most other materials). Thus, at 300°K, we might observe a value for $(1/\sigma)(d\sigma/dT)$ as high as about 4% degK$^{-1}$.

In practice, thermistors are not made from pure semiconductors, although germanium resistance thermometers have proved useful at low temperatures (Blakemore, et al., 1962) (the germanium is not, of course, then an intrinsic conductor). Most thermistors use oxide semiconductors, the oxides nowadays usually being those of the transition metals (Scarr and Setterington, 1960). Such materials are far less sensitive to variations in the concentration of impurities than are the semiconductors that one uses in junction devices, and they can be prepared by sintering the compacted powders.

In an attempt to understand the conduction mechanisms that operate for the transition metal oxides, a detailed study has been made of nickel oxide, NiO. When pure and stoichiometric, NiO is an insulator, but its conductivity can be raised to values of the order of 100 $\Omega^{-1}$ $m^{-1}$ by replacing some of the nickel atoms by lithium atoms to form $Li_x Ni_{1-x}O$, where $x$ is of the order of 1 at.%. Such a concentration of lithium is, of course, many orders of magnitude greater than that of the impurities in transistor-grade materials. The effect of the substitution of monovalent lithium (or alternatively of non-stoichiometry) is to cause some of the normal $Ni^{2+}$ ions to be replaced by $Ni^{3+}$ ions. It is the extra electrons associated with the $Ni^{3+}$ ions that are responsible for the semiconducting behaviour.

In NiO the conduction electrons must be regarded as lying in very narrow energy bands with correspondingly large effective masses (hence the observation of semiconducting behaviour for such large concentrations of doping agent). Generally, the electrical conductivity rises with temperature according to an exponential law but there has been some doubt as to whether this behaviour should be attributed to an increase of the carrier concentration or of the mobility as the temperature is raised (Austin et al., 1967). The most popular theory (Heikes and Johnston, 1957) attributes the fact that the conductivity is proportional to $\exp(-A/kT)$, where $A$ is some activation energy, to the hopping of electrons from site to site. The activation energy then represents the height of the barriers between the sites.

The earliest experiments on NiO were performed using the same sort of polycrystalline sintered samples that are employed in thermistors, and it was at first thought that the grain boundaries might be having a dominating influence. However, more recent studies have been made on single crystals and the exponential increase of conductivity with temperature is still observed.

When thermistors are used in the measurement of temperature, one normally requires a rapid but smooth variation of resistance with temperature over a reasonably wide range. However, there are a number of applications of thermistors that make use of the non-linear variation of voltage drop with current. The electrical resistance of a thermistor falls with increasing current owing to the Joule heating effect. In fact, for a given thermistor, working at a given ambient temperature, the plot of voltage against current displays a maximum as shown in Figure 1.1. Thus, a thermistor can be used in electrical control systems, for example as a voltage-stabilising element. Most thermistors have a relatively high thermal inertia and can be used in surge suppression. However, for other applications a much shorter time constant would be desirable. For example, the voltage–current plot suggests that a thermistor might be used as a two-state logic element but a fast response as well as a more pronounced negative resistance characteristic would be needed.

Recently a new type of thermistor has become available that meets these requirements. Morin (1959) found that the single crystals of various oxides of vanadium all displayed discontinuities in their conductivity versus temperature plots. These discontinuities, amounting sometimes to several orders of magnitude in the conductivity, occurred in the region of a temperature $T_c$, the material being antiferromagnetic below $T_c$ and paramagnetic above $T_c$. Although the change in conductivity at $T_c$ is only of about two orders of magnitude for the oxide $VO_2$, this material is particularly useful since $T_c$ is about 68°C. Figure 1.2 shows the variation of electrical resistance with temperature for a sintered bead of $VO_2$ compared with that for an ordinary thermistor (Futaki, 1965). Not surprisingly, the fall in resistance of a $VO_2$ thermistor, when the current exceeds the value at the voltage maximum, is much more pronounced than for an ordinary thermistor. It is also found that samples of $VO_2$ can be switched from a high-resistance state to a low-resistance state (and vice versa), in time intervals of the order of 1 $\mu$s, by short pulses (Cope and Penn, 1968). The short response time is due to the fact that switching involves only the formation or rupture of a very thin filament or neck of high conductivity $VO_2$ $(T > T_c)$ within low-conductivity material $(T < T_c)$.

The behaviour of $VO_2$ (and of the other oxides of vanadium) can be understood in simple terms if it is supposed that, in the antiferromagnetic state, there are two narrow bands separated by an appreciable energy gap. In a pure stoichiometric sample, at 0°K, the upper band would be empty and the lower band full. In real samples we can expect carriers to be present in one of the bands due to non-stoichiometry and, below $T_c$, semiconducting characteristics are apparent. Above $T_c$, when the oxide is no longer antiferromagnetic, the two bands become one which can be

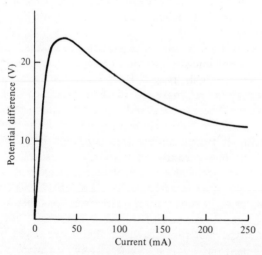

**Figure 1.1.** Typical voltage–current characteristic for a thermistor.

no more than about half-filled. This is the condition for metallic conduction to appear and, as observed by Morin, the electrical conductivity falls with rise of temperature if $T$ is appreciably greater than $T_c$.

Futaki (1965) has listed a large number of possible applications for the $VO_2$ type of thermistor, which he calls a critical-temperature resistor.

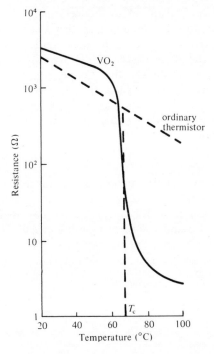

**Figure 1.2.** Plot of resistance against temperature for a $VO_2$ bead (after Futaki, 1965).

### 1.2 Thermoelectric energy convertors

When an electric current is passed through a conductor, there is a transport of heat, which is observable as the Peltier effect at the junction between two such conductors. Because the energy flows on either side of the junction are generally different, there is a liberation or absorption of heat to or from the surroundings.

Similarly, a temperature gradient applied to a conductor produces a potential gradient, which is revealed when different conductors are joined together; this phenomenon is the Seebeck effect. Kelvin showed that the Peltier coefficient $\pi$ and the Seebeck coefficient $\alpha$ are related by the simple thermodynamic rule

$$\pi = \alpha T \,. \tag{1.4}$$

Since their discovery, early in the nineteenth century, it has been clear that the thermoelectric effects could be used either in the generation of electricity from heat or in refrigeration. However, it is only since the advent of semiconductors with controlled properties that reasonably efficient devices based on the Seebeck and Peltier effects have become available.

Qualitatively, it is easily seen that a good thermoelectric material should have a high Seebeck (or Peltier) coefficient. Also the electrical conductivity should be high so as to minimise Joule heating losses, and the thermal conductivity should be low so that losses by heat conduction are reduced. Early in the century, Altenkirch (1909, 1911) expressed these considerations in quantitative form; he showed that the performance of a well-designed thermocouple can be expressed solely in terms of the temperatures of the hot and cold junctions and a quantity, known as the figure of merit $z$. The figure of merit is defined by the equation

$$z = \frac{(\alpha_p - \alpha_n)^2}{[(\kappa_p/\sigma_p)^{\frac{1}{2}} + (\kappa_n/\sigma_n)^{\frac{1}{2}}]^2} \,, \tag{1.5}$$

where $\kappa$ is the thermal conductivity and the subscripts p and n refer to the positive and negative branches respectively. $\alpha_p$ and $\alpha_n$ are the so-called absolute Seebeck coefficients measured with reference to an ideal metal (Christian et al., 1958). Figures 1.3 and 1.4 show how the efficiency of a thermoelectric generator and the coefficient of performance [1] of a thermoelectric refrigerator depend on the ratio of cold-junction

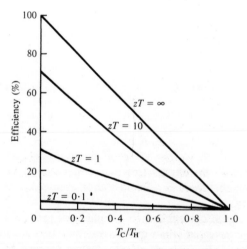

**Figure 1.3.** Plot of efficiency against ratio of cold- to hot-junction temperature for thermoelectric generators having various values of the dimensionless figure of merit $zT$.

[1] The coefficient of performance of a refrigerator is the ratio of the cooling capacity to the electrical power consumption.

temperature $T_C$ to hot-junction temperature $T_H$ for various values of $zT$, the dimensionless figure of merit. Worthwhile values of the generator efficiency can be reached only if $zT$ is of the order of unity or greater. Similarly, a reasonably large depression of cold-junction temperature by a thermoelectric refrigerator (the coefficient of performance remaining positive) can be found only if the same condition holds good.

Actually, in searching for thermoelectric materials, one rarely makes use of the figure of merit as defined by Equation (1.5). Instead one uses a figure of merit for a single material defined by

$$z = \frac{\alpha^2 \sigma}{\kappa}.$$

(1.6)

The true figure of merit for a couple is usually very close to the average of the values of $z$, as defined by Equation (1.6), for each of its branches.

It was natural enough, at first, to attempt to use metals and metallic alloys in thermoelectric generators and refrigerators. However, it is now realised that metals never can have large enough values of $zT$. The ratio of thermal to electrical conductivity is the same for all metals at a given temperature and, according to the Wiedemann-Franz-Lorenz law, this ratio is equal to $(\pi^2/3)(k/e)^2 T$. The Wiedemann-Franz law is a consequence of the fact that the conduction electrons in a metal carry both the electrical and thermal currents. If the ratio $\kappa/\sigma$ is given by $(\pi^2/3)(k/e)^2 T$, it is readily shown that $zT$ can reach unity only if $|\alpha|$ is about 160 $\mu$V degK$^{-1}$. Both theory and experiment indicate that such a high Seebeck coefficient cannot be achieved for a degenerate conductor, i.e. a metal.

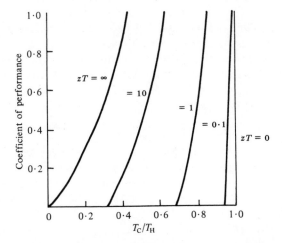

**Figure 1.4.** Plot of coefficient of performance against ratio of cold- to hot-junction temperature for thermoelectric refrigerators with different values of $zT$.

Semiconductors may have much larger values of the Seebeck coefficient, reaching, for some samples, several mV degK$^{-1}$. However, such values are always accompanied by very small electrical conductivities. The contribution of the charge carriers to the thermal conductivity is certainly very low too, but the total thermal conductivity can still be quite high. The reduction of the electronic contribution allows the heat conductivity due to the lattice vibrations to predominate. In other words, if we write the total thermal conductivity as

$$\kappa = \kappa_L + \kappa_e \, , \tag{1.7}$$

where $\kappa_L$ is the lattice component and $\kappa_e$ the electronic component, then, for most semiconductors, $\kappa_L \gg \kappa_e$.

In fact, it turns out that semiconductors with very large Seebeck coefficients are not much use in thermoelectric energy convertors, since the high values of $|\alpha|$ are more than offset by the very low value of $\sigma/\kappa$. It is much better to add impurities to the semiconductor so that $|\alpha|$ is only about a couple of hundred $\mu$V degK$^{-1}$. This is easily demonstrated theoretically if we make use of the classical expressions for the various quantities involved in the figure of merit (Ioffe, 1957; Goldsmid, 1960). Then, for an extrinsic semiconductor

$$\alpha = \pm \frac{k}{e} (\lambda + \tfrac{5}{2} - \eta) \, , \tag{1.8}$$

where the upper sign refers to a p-type conductor and the lower sign to an n-type conductor. It is supposed that the relaxation time for the charge carriers is proportional to $E^\lambda$, where $E$ is the energy and $\lambda$ is a constant for a given scattering mechanism. $\eta$ is the reduced Fermi energy (i.e. the Fermi energy divided by $kT$), which is positive if the Fermi level lies within the appropriate energy band and negative if it lies in the forbidden gap. One can regard $(\lambda + \tfrac{5}{2})kT$ as the average kinetic energy carried by the electrons or holes and $-\eta kT$ as their potential energy, so that the Seebeck coefficient is proportional to the sum of the kinetic and potential energy of the carriers.

The electrical conductivity is given by

$$\sigma = ne\mu \, , \tag{1.9}$$

where the mobility $\mu$ is supposed to be independent of the carrier concentration $n$. We can regard the Fermi energy of a semiconductor as an adjustable parameter since it is related to the carrier concentration by the expression

$$n = 2 \left( \frac{2\pi m * kT}{h^2} \right)^{3/2} \exp(\eta) \, . \tag{1.10}$$

The carrier concentration can, of course, be changed by doping the material.

Finally, for a non-degenerate semiconductor the electronic thermal conductivity is

$$\kappa_e = (\lambda + \tfrac{5}{2})\left(\frac{k}{e}\right)^2 \sigma T, \tag{1.11}$$

while the lattice component is supposed to be substantially independent of the carrier concentration.

We make use of Equations (1.8) to (1.11) to express the figure of merit in terms of the reduced Fermi energy:

$$zT = \frac{\{\eta - (\lambda + \tfrac{5}{2})\}^2}{1/\beta \exp(\eta) + (\lambda + \tfrac{5}{2})} \tag{1.12}$$

where

$$\beta = 2\frac{k^2}{e}\left(\frac{2\pi m k}{h^2}\right)^{3/2} \frac{\mu}{\kappa_L}\left(\frac{m*}{m}\right)^{3/2} T^{5/2}.$$

The optimum value for the Fermi energy is found by setting $\mathrm{d}(zT)/\mathrm{d}\eta$ equal to zero, whence

$$\eta_{\mathrm{opt}} + 2(\lambda + \tfrac{5}{2})\beta\exp(\eta_{\mathrm{opt}}) = \lambda + \tfrac{1}{2}. \tag{1.13}$$

Actually, for all reasonable values of $\lambda$ (ranging from $-\tfrac{1}{2}$ for acoustic mode lattice scattering to $+\tfrac{3}{2}$ for simple ionised-impurity scattering) and of $\beta$, it is found that $\eta_{\mathrm{opt}} > -2$, which means that classical statistics cannot be regarded as a good approximation. However, if one performs an exact calculation using Fermi-Dirac statistics, it still turns out that $\eta_{\mathrm{opt}}$ has more or less the same value, that is about zero; the optimum Seebeck coefficient is then approximately $\pm200$ $\mu$V degK$^{-1}$. Furthermore, the prediction from Equation (1.12) that, for a given Fermi energy, the figure of merit rises with increasing $\beta$, is valid. In other words, the product of mobility and effective mass raised to the power $\tfrac{3}{2}$, for the majority carriers, should be high, and the lattice thermal conductivity should be low.

There is another condition that must be satisfied if the above considerations are to hold good. The energy gap must be great enough for a Seebeck coefficient of about $\pm200$ $\mu$V degK$^{-1}$ to be achieved with a negligible concentration of minority carriers. The contributions to the Seebeck and Peltier effect from electrons and holes are in the opposite sense, so it is unfavourable if both types of carrier make a significant contribution to the conductivity. Moreover, it is found that electron-hole pairs in a mixed conductor carry their ionisation energy down a temperature gradient, thus enhancing the thermal conductivity. It is necessary, then, that $E_g \gg kT$.

Semiconductors with a large effective mass tend to have a low mobility, while those with a small effective mass usually have a high mobility. However, provided that intra-valley scattering is dominant, one can obtain reasonably large values of both the mobility and the effective mass by choosing a multi-valley semiconductor; in such a material the inertial effective mass that determines the carrier mobility can be low, but the overall density of states mass is $N_v^{2/3}$ times that for a single valley, where $N_v$ is the number of equivalent energy minima. Otherwise it is not too easy to predict the materials that will have high values of $\mu(m*/m)^{1/2}$, except that they should not be strongly ionic compounds.

It is rather easier to forecast those materials that will have a low lattice thermal conductivity. Except at low temperatures, the scattering of phonons in pure elements and compounds is due primarily to collisions with other phonons. It is expected that, at a given temperature, the lattice vibrations will be strongest (and the phonon scattering most intense) for materials with low Debye temperatures and low melting points. Such materials are those of high atomic weight (or mean atomic weight). Furthermore, the lattice thermal conductivity of a given element or compound can be reduced by forming a solid solution between it and an isomorphous substance. Rather surprisingly this does not mean that the mobility of the charge carriers is also reduced. This is because the wavelengths associated with the charge carriers in a semiconductor are much greater than the interatomic spacing, whereas the heat-conduction phonons have quite short wavelengths. The short-range disorder in the solid solution scatters the shorter waves much more strongly than the longer ones.

It is concluded that a good thermoelectric material is likely to be a many-valley semiconducting solid solution formed from elements of high atomic weight, or compounds of such elements. The alloys based on the compound bismuth telluride $Bi_2Te_3$, which are commonly used in thermoelectric refrigerators, exemplify these principles and it is instructive to examine their behaviour.

Figure 1.5 shows the Seebeck coefficient plotted against electrical conductivity for p-type and n-type samples doped to various levels. Samples displaying mixed conduction (represented on the left-hand side of the figure) have low Seebeck coefficients as well as low electrical conductivities. The more useful materials are extrinsic conductors for which, as expected, $|\alpha|$ falls as $\sigma$ rises. The solid curve corresponds to the compound $Bi_2Te_3$ while the dashed curves represent solid solutions of $Bi_2Te_3$ with $Sb_2Te_3$ and $Bi_2Se_3$. The fact that the dashed and solid curves lie so close to each other is an indication that there can be very little difference between the carrier mobilities in the compound and in its alloys.

Figure 1.6 shows how the thermal conductivity varies with electrical conductivity for the same materials. The relatively large values of thermal conductivity on the left-hand side of the figure are due to bipolar heat flow in the mixed conduction region. In the extrinsic region the electronic thermal conductivity rises with electrical conductivity. The

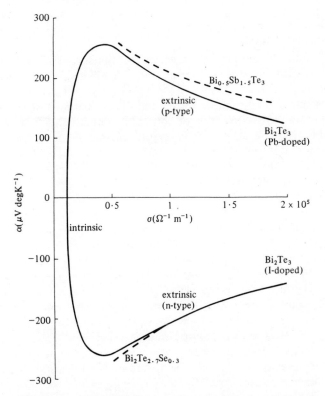

**Figure 1.5.** Plot of Seebeck coefficient against electrical conductivity for bismuth telluride and its alloys.

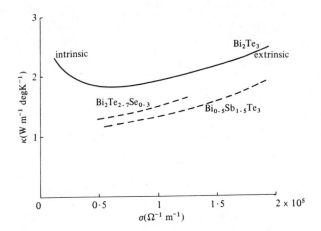

**Figure 1.6.** Plot of thermal conductivity against electrical conductivity for bismuth telluride and its alloys.

observation that the curves for the solid solutions are appreciably lower than that for $Bi_2Te_3$ illustrates the reduction of lattice conductivity due to alloy scattering of phonons.

Figure 1.7 indicates the manner in which the dimensionless figure of merit $zT$ varies with Seebeck coefficient and, as predicted, it reaches its maximum at $|\alpha| \approx 200\ \mu V\ degK^{-1}$. It is noteworthy that $Bi_2Te_3$ is a multi-valley semiconductor (there are six equivalent energy minima for both electrons and holes) and it is, of course, a compound of high mean atomic weight. Actually it has an energy gap of only about $5kT$ at room temperature and, at elevated temperatures, it becomes impossible to achieve a Seebeck coefficient as high as $200\ \mu V\ degK^{-1}$. For this reason, other thermoelectric materials become superior to $Bi_2Te_3$ above about $200°C$. Some of the more important materials are shown in the plot of $zT$ against temperature (Figure 1.8).

Although the figures of merit that have been realised are not high enough for thermoelectric generators and refrigerators to oust more conventional techniques of energy conversion from most of the fields of application, there are several examples where their employment has been worthwhile. The most important use of thermoelectric generators has been in power sources for remote locations; using a radio-isotope source such as $^{90}Sr$ there is negligible fall-off in output over a period of ten years or more. Thermoelectric refrigerators are used when relatively small amounts of cooling power are needed. They have found widespread applications in the cooling of electronic apparatus and scientific equipment, the ease of control of the output being an important advantage. It should be noted that most applications require a thermo-electric module consisting of a number of thermocouples connected electrically in series; by this means the required cooling power can be achieved with a relatively small current flow.

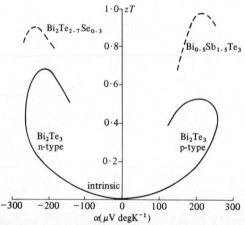

**Figure 1.7.** Plot of $zT$ against Seebeck coefficient of bismuth telluride and its alloys.

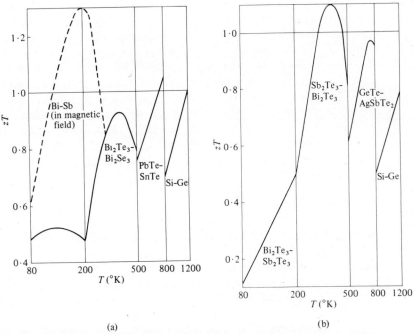

**Figure 1.8.** Plot of $zT$ against temperature for some of the best thermoelectric materials: (a) n-type; (b) p-type.

## 1.3 Hall effect and magnetoresistance devices

### 1.3.1 Applications of the Hall effect

The Hall coefficient $R_H$ of an extrinsic semiconductor indicates the sign of the charge carriers and their concentration. Generally,

$$R_H = \mp \frac{A}{ne} , \qquad (1.14)$$

where the negative sign is applicable for electrons and the positive sign for holes. $A$ is a number that is close to unity but which depends for its precise value on the carrier scattering mechanism and on the degree of degeneracy (Putley, 1960). Only rarely does $A$ differ from unity by more than a factor of two; for a non-degenerate conductor, in which acoustic-mode lattice scattering is dominant, $A$ is equal to $3\pi/8$, that is $1\cdot18$.

The Hall coefficient is defined by the relation

$$\mathcal{E}_y = R_H j_x B_z , \qquad (1.15)$$

where $j_x$ is the current density in the $x$ direction, $B_z$ the magnetic field in the $z$ direction, and $\mathcal{E}_y$ the resultant electric field (the Hall field) in the $y$ direction. Thus, for a material in which the Hall coefficient has been established, the measurement of $\mathcal{E}_y$ provides a means of determining the product $j_x B_z$.

It is generally the aim in designing a Hall-effect device to produce the maximum Hall voltage for a given magnetic field $B_z$. At first sight it might appear that this simply means that the carrier concentration should be as small as possible. However, it turns out that the mobility of the charge carriers is an even more important quantity than the carrier concentration. Consider the circuit shown in Figure 1.9. A current $I_x$ is passed along the Hall element and the Hall probes are connected to a load of resistance $r_L$. The ratio of the power that is delivered to the load to the input power (i.e. the transfer efficiency) is highest when the load resistance is equal to the transverse resistance of the Hall element. Under this condition it can be shown that

$$\text{transfer efficiency} = \frac{B_z^2 R_H^2 \sigma^2}{4f_1} \, , \tag{1.16}$$

when $f_1$ is a geometric factor that takes account of end effects. For a given magnetic field, the transfer efficiency depends only on the product $R_H \sigma$, which is known as the Hall mobility $\mu_H$.

Of course, it is not often that one delivers the output from a Hall element into a matched load. A common situation is that in which the output is fed into an amplifier with an input impedance far in excess of that of the Hall element. One is then interested in obtaining the largest open-circuit Hall voltage $V_y$ for a given input power $W_i$. It is found that

$$V_y = W_i^{\frac{1}{2}} \left(\frac{d}{tl}\right)^{\frac{1}{2}} f_2 B_z R_H \sigma^{\frac{1}{2}} \, , \tag{1.17}$$

where $l$ is the length of the element, $d$ its width between the Hall probes, $t$ its thickness, and $f_2$ is another geometric factor. Equation (1.17) shows that in this case the product of the Hall mobility $\mu_H$ and the Hall coefficient $R_H$ should be as large as possible. In other words one would like a small carrier concentration as well as a high mobility. Actually, a small carrier concentration is desirable even if one is operating the device

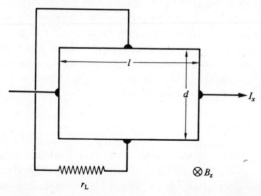

**Figure 1.9.** Diagram showing a load resistance connected across a Hall element.

with maximum transfer efficiency, since it is difficult to achieve the matched load condition if the Hall element is made from a material of very low electrical resistivity.

The various materials that might be used in Hall-effect devices have been discussed by Hilsum (1961). Germanium can be prepared with a very low carrier concentration and, thus, with a large Hall coefficient. Of course, if the material is intrinsic, the Hall coefficient will be strongly temperature-dependent but, unless the energy gap is very small, relatively light doping will reduce this temperature dependence considerably. For example, an increase of the electron concentration from the intrinsic value of $2 \times 10^{19}$ m$^{-3}$ to $2 \times 10^{21}$ m$^{-3}$ reduces the (negative) temperature dependence of the Hall coefficient from about 5% degK$^{-1}$ to 0·01% degK$^{-1}$.

Germanium is actually unsuitable for most Hall effect applications as its electron mobility is only about 0·4 m$^2$ V$^{-1}$ s$^{-1}$ even when it is pure. Much higher electron mobilities can be achieved using some of the compound semiconductors. Thus, the mobility of electrons in indium antimonide, InSb, can be as large as about 8 m$^2$ V$^{-1}$ s$^{-1}$ and this compound is generally regarded as the best material for most galvano-magnetic devices.

A disadvantage of InSb is its rather small energy gap, about 0·2 eV compared with about 0·7 eV for germanium. This means that even the intrinsic material has a rather low electrical resistivity. It is interesting that the electrical resistivity can be raised, while the temperature dependence of the Hall coefficient is reduced to almost zero over a limited temperature range, by doping InSb with an *acceptor* impurity so that the hole concentration becomes an order of magnitude greater than the electron concentration. The material still appears to be an n-type conductor since the electron mobility (which is about 4 m$^2$ V$^{-1}$ s$^{-1}$ for the doped material) is so much greater than the hole mobility (about 0·06 m$^2$ V$^{-1}$ s$^{-1}$). The Hall coefficient of a mixed semiconductor is given by (Putley, 1960)

$$R_H \approx \frac{p\mu_p^2 - n\mu_n^2}{e(p\mu_p + n\mu_n)^2} , \tag{1.18}$$

where $n$ is the electron concentration and $p$ the hole concentration. Clearly the Hall coefficient remains negative so long as $p/n < (\mu_n/\mu_p)^2$.

The rise of resistivity on doping InSb with an acceptor is due, of course, to the fact that the electron concentration is reduced, since the product $np$ must remain almost constant[2]. However, even for acceptor-doped material the resistivity does not rise much above $10^{-4}$ Ω m. This means that Hall elements with acceptably high resistances for matching to conventional electronic equipment (i.e. ohms rather than small fractions of an ohm) can be made from InSb only if the samples are very thin. Ideally

---

[2] $np$ is precisely constant only if classical statistics are applicable.

the material should be in the form of a film but, until recently, films of InSb invariably had much lower electron mobilities than the bulk material. However, it is now possible to produce thin films of high-mobility InSb by a recrystallisation technique (Carroll and Spivak, 1966; Wieder, 1966); the films must be at least 2 or 3 $\mu$m in thickness if boundary scattering at the surfaces is not to reduce the mobility.

Hall elements are commonly used for the measurement of magnetic fields. In most cases the aim is to measure fields of the order of $0 \cdot 01$ T upwards but by using permalloy flux concentrators and by amplifying the Hall voltage, fields as low as $10^{-10}$ $T$ can be detected (Hilsum, 1961).

If the magnetic field is produced by an electric current, then the output of the Hall element is a function of this current and can be used to

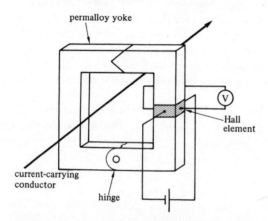

**Figure 1.10.** Clip-on ammeter using a Hall-effect detector.

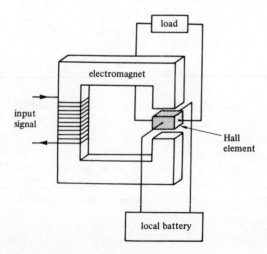

**Figure 1.11.** Hall-effect amplifier.

measure it. The simplest device based on this principle is a clip-on ammeter which incorporates a Hall element as the detector, as shown in Figure 1.10. The same idea has been used in the measurement of electron and ion beam currents down to about 5 $\mu$A: in detecting such small currents the Hall element has been fed with a.c. and the output amplified (Whitlock and Hilsum, 1960).

Hall elements can be used as analogue multipliers, wattmeters, modulators, and in the performance of a great variety of other electronic functions. One of the most interesting applications is the low-noise amplifier illustrated in Figure 1.11. The input signal is applied to the coil of an electromagnet while the Hall element is supplied from a local battery. Using an InSb Hall-effect amplifier at room temperature, Ross and Thompson (1955) have obtained a power gain of 5.

### 1.3.2 Magnetoresistance

The application of a magnetic field to a current-carrying conductor generally not only produces a transverse (Hall) electric field but also changes the longitudinal electric field. In other words the resistance of the conductor is altered. The magnetoresistance effect (which nearly always manifests itself as an increase in resistance) arises from the fact that, owing to the Lorentz force, the charge carriers have a transverse component of their drift velocity. If all the carriers had the same relaxation time, this transverse component would be zero under the condition of zero transverse current; the Lorentz force would be exactly balanced by the force due to the Hall field. However, in practice, there will always be a transverse flow in one direction for some of the carriers which will be balanced by a transverse flow in the opposite direction for the remaining carriers. These transverse flows will be enhanced by a strong energy dependence of the relaxation time and by any complexities of the energy band structure but, to obtain the largest magnetoresistance effect, it is necessary to prevent the establishment of a transverse electric field.

There would be only a very small transverse electric field if the sample were an intrinsic semiconductor having electrons and holes of comparable mobility. The electrons and holes could flow together in a transverse direction without transporting any net charge. It is, however, very difficult to find semiconductors having large mobilities for both the electrons and holes. Thus, one usually has to destroy the transverse electric field by some other means.

Some of the methods of allowing transverse current flow in the semi-conductor, thereby increasing the magnetoresistance effect, are shown in Figure 1.12. In the Corbino disc arrangement the electrodes are connected at the centre and the circumference of a circular disc of material and the magnetic field is applied perpendicular to the plane surfaces. The magneto-resistance is also large for a short, wide sample because the electrodes effectively short-circuit the Hall voltage. Both the Corbino disc and the short sample tend to suffer from the disadvantage (for most applications) of a rather low electrical resistance. This disadvantage can be overcome by connecting several short samples in series, a configuration that is achieved by means of transverse metal strips applied to a long sample as in Figure 1.12c.

One of the most elegant methods of obtaining the raster-bar arrangement of Figure 1.12c has been described by Weiss (1966). He has grown material of the composition InSb + 1·8% NiSb, corresponding to the eutectic between the semiconductor InSb and the metal alloy NiSb. The NiSb takes the form of fine needles which lie parallel to one another within the InSb; these needles reduce the Hall effect when the directions of the applied electric current and magnetic field lie perpendicular to them. Using his eutectic material, Weiss has observed a fifteenfold increase of resistance over the zero-field value on applying a magnetic field of 1 T.

Many devices could make use of either the Hall effect or the magneto-resistance effect. Which effect should one choose? The Hall effect is directly proportional to the magnetic field while the magnetoresistance effect depends on the square of the field. Thus as a general rule one expects the Hall effect to be superior at low fields and the magneto-resistance effect to become the better at high fields. The value of the field at which the changeover in preference occurs is given by $\mu B = 1$. The magnetoresistance effect for zero transverse electric field is given

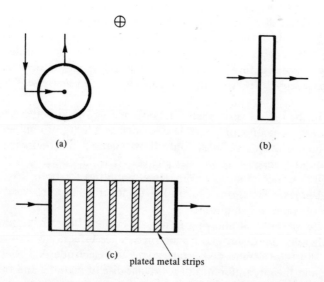

(a)

(b)

(c)

plated metal strips

**Figure 1.12.** Methods for removing the transverse electric field in a magnetoresistance element: (a) Corbino disc; (b) short sample; (c) raster-bar arrangement.

approximately by

$$\rho(B) = \rho(0)(1+\mu^2 B^2),$$ (1.19)

where $\rho(B)$ is the resistivity in a field $B$ and $\rho(0)$ the resistivity in zero field.

Perhaps the most promising use of magnetoresistance elements is in variable resistors that do not require sliding contacts (contactless potentiometers). A good application, described by Ross and Saker (1957), is the measurement of very small displacements using the arrangement shown in Figure 1.13. The transducer consists of a hollow rectangular sheet of InSb with copper-plated contacts that divide it into four sections, A, B, C, and D. The centre of one pair of sections A-B lies between the poles of a 1 T permanent magnet so that these sections display equal magnetoresistance effects; the resistance bridge remains balanced. However, a small displacement upwards or downwards with respect to the magnet will increase the magnetoresistance in one section and decrease it in the other. The displacement is thus revealed by an out-of-balance signal. Hilsum (1961) has reported the detection of movements of about 5 Å using such a device.

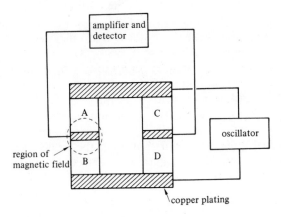

**Figure 1.13.** Magnetoresistance displacement transducer.

## 1.4 The Gunn effect

When a high electric field is applied to a semiconductor it is commonly observed that Ohm's law breaks down. This is because the field imparts to the charge carriers an amount of energy that is comparable with or larger than their thermal energy. When the carriers become 'hot', their relaxation time is usually altered and the mobility no longer has the low-field value. The non-ohmic behaviour associated with the heating of the carriers is observed more or less instantaneously on the application of the field, since the lattice remains at its original temperature.

Particularly interesting effects occur when the departures from Ohm's law lead to a negative-resistance region in the current-voltage characteristic, especially if this negative resistance is of the *voltage-controlled* type, in which the current is a single-valued function of the voltage (cf. Figure 1.1 which shows a *current-controlled* negative resistance for a thermistor, where the voltage is a single-valued function of the current). One semiconductor device that displays a voltage-controlled negative resistance is the tunnel diode (to be described later); here we are concerned with the rather different behaviour when voltage-controlled negative resistance appears in a bulk semiconductor rather than at a junction (Bott, 1968; Butcher, 1967).

Figure 1.14 shows a schematic plot of electron velocity against electric field for an n-type sample of the compound gallium arsenide, GaAs. The velocity of the electrons, and hence the current, increases with the field up to a value of about $2 \times 10^5$ m s$^{-1}$ at $3 \times 10^5$ V m$^{-1}$ but at higher fields the velocity falls to give a negative differential resistance. This behaviour can be understood from a consideration of the energy-band structure of GaAs which is illustrated in Figure 1.15. The minimum conduction band energy is found at the zero of wavevector space (centre of the Brillouin zone) but there are minima for satellite valleys along the $\langle 100 \rangle$ directions in the Brillouin zone, that have an energy that is $0 \cdot 36$ eV higher than that of the central minimum. The electrons in the central valley have a low effective mass and a high mobility, $0 \cdot 75$ m$^2$ V$^{-1}$ s$^{-1}$. On the other hand, in the satellite valleys the effective mass is comparatively high and the mobility is only about $0 \cdot 015$ m$^2$ V$^{-1}$ s$^{-1}$. At low electric fields, the electrons are all to be found near the bottom of the central valley, but at high fields the energy of the electrons rises considerably. When the increase of energy amounts to about $0 \cdot 36$ eV, the electrons are easily scattered by longitudinal optical-mode phonons into the satellite valleys which have a high density of states. It is this transfer of the electrons into

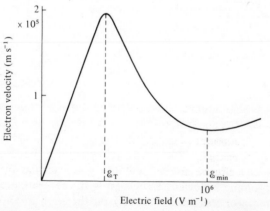

**Figure 1.14.** Typical plot of electron velocity against electric field for n-type GaAs.

the low-mobility satellite valleys that accounts for the fall in electron velocity above the threshold field $\mathcal{E}_T$.

At first sight it might appear that the current-voltage characteristic of a sample of GaAs would be of the same form as the velocity-field plot shown in Figure 1.14. However, the situation in which the whole sample is at a uniform field in the negative-resistance region is unstable. The potential gradient becomes non-uniform and one sees the formation of a 'domain', in which the field is very high, while the remainder of the sample is subject to a field $\mathcal{E}_0$ that lies below the threshold value. Figure 1.16 shows how the field and electron density vary with position for a sample that is long compared with the domain length; it has been found that for GaAs this means that the product of electron concentration

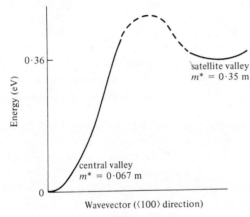

**Figure 1.15.** Conduction band structure shown schematically for GaAs.

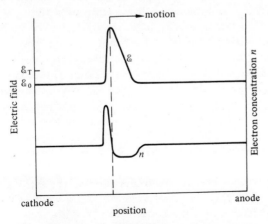

**Figure 1.16.** Electric field and electron density plotted against position for a sample of GaAs showing a fully formed domain. The domain moves from cathode to anode with the drift velocity of the electrons in the field $\mathcal{E}_0$.

and length must be appreciably greater than $10^{16}$ m$^{-2}$. Under this condition the shape of the domain is stable but its position changes with time; it moves along the sample with the same velocity as that of the electrons in the field $\mathcal{E}_0$. Since the velocity of electrons in the high-field region of the domain is less than that of the electrons outside the domain, it is clear that the negative charge concentration must be higher than average near the trailing edge and lower than average near the leading edge. In other words a dipole is superimposed on the otherwise uniform charge distribution.

The domain is nucleated in the region of the cathode and, after formation, moves to the anode where it disappears. When it disappears the current rises, falling again when the next domain is formed. This, then, is the origin of the current oscillations that constitute the Gunn effect (Gunn, 1963). The current is modulated with a frequency that depends on the time of transit of domains. If $l$ is the length of the sample and $v_D$ is the drift velocity of the electrons, the transit line is $l/v_D$. Thus, for a sample that is 100 $\mu$m long, Figure 1.14 shows that the transit time will be rather less than $10^{-9}$ s and the current will be modulated at rather more than 1 GHz. The possibility of a solid-state microwave generator is immediately apparent.

Actually, the 'transit' mode of operation that has been described is only one of a number of ways utilising the Gunn effect. It suffers from the disadvantage that the frequency is fixed by the length of the sample. We, therefore, discuss some of the other modes of operation.

Suppose that the sample is part of a resonant circuit. In this case it is expected that the voltage across the sample will no longer be constant but it will tend to be modulated at the resonant frequency. It is then possible for the field to lie below the threshold value at the time when the domain disappears at the anode; a new domain will not form until the field rises above $\mathcal{E}_T$ once more. In this 'delayed' mode the frequency of the current pulses is reduced below that $f_T$ for the transit mode. Alternatively, it is possible to arrange that the field falls below the value that is needed to sustain the domain before the latter reaches the anode. The original domain is quenched and a new domain is formed as soon as the field rises to the threshold value. Under this mode of operation one can achieve a frequency higher than $f_T$. It is thus possible to vary the frequency of operation of a Gunn device of given length, but the frequencies that can be covered in this way would seem to be restricted to the range $f_T/2$ to $2f_T$.

Perhaps most interesting is the so-called LSA (limited space-charge accumulation) mode. The time for the formation of a domain is of the order of the dielectric relaxation time $\tau_r$ equal to $\epsilon/\sigma$, where $\epsilon$ is the permittivity. Suppose, however, that the field is modulated at a frequency that is much greater than $1/\tau_r$ and that, as shown in Figure 1.17, it falls below $\mathcal{E}_T$ over part of each cycle. In this situation the domains do not have time to form properly. The space-charge fluctuations grow while the field lies between $\mathscr{E}_T$ and $\mathscr{E}_{\min}$ and decay while the field lies below $\mathscr{E}_T$ [3]. The period of time for this decay must be large enough for cumulative growth of the domains to be prevented. High efficiencies for operation in the LSA mode are achieved using a large d.c. field with a large a.c. field superimposed upon it.

It is obvious that there is a minimum frequency of operation in the LSA mode for a given value of $\tau_r$. The dielectric relaxation time can be increased by reducing the carrier concentration $n$ but it is found that the efficiency for a given frequency falls if $n$ is made too small. In practice, the LSA mode can be usefully employed if the ratio of carrier concentration to frequency lies between $2 \times 10^{10}$ and $10^{11}$ s m$^{-3}$. It has been used successfully up to a frequency of about $10^{11}$ Hz. In effect the sample displays its static characteristic (Figure 1.14) and behaves as a negative-resistance amplifier or oscillator in the same way as does a tunnel diode.

Next, let us briefly consider the movement of a domain in the transit mode through a sample in which the cross-section area or doping level changes over the length. Since the velocity of the domain is almost independent of the local electric field, the current is proportional to the product of carrier concentration and area $A$, at the domain position. To a first approximation the carrier concentration is equal to the donor concentration $N_D$, so the current varies as the product $N_D(x)A(x)$, where $x$ locates the domain. Thus, the current can be modulated by deliberately changing either the area or the impurity concentration along the sample.

There is a whole new class of interesting devices with non-uniform area

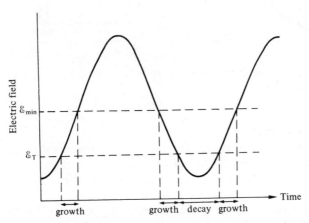

**Figure 1.17.** Electric field against time for a Gunn device operated in the LSA mode with a large d.c. bias.

[3] Growth and decay can be neglected for $\mathscr{E} > \mathscr{E}_{\min}$ since the slope of velocity-field plot is then close to zero.

or doping level. For example, if the sample of GaAs takes the form of a disc, as shown in Figure 1.18a, with the cathode at the centre and the anode at the circumference, it behaves as a tunable Gunn oscillator. The electric field is strongest near the centre of the disc and weakest near the circumference (quite apart from the field modifications induced by the formation of a domain). Thus it is possible for a domain to be formed near the cathode and for it to move outwards into a region of decreasing electric field until it is eventually quenched. The higher the applied voltage, the further will the domain travel before quenching occurs, and the lower will be the frequency of oscillation.

A similar device is the voltage-controlled frequency switch shown in Figure 1.18b. For a moderate applied voltage, domains can be formed at the cathode, but quenched just before they reach the centre of the sample because the field is weakest in this region. At a higher applied voltage, the field in the central region will be strong enough to sustain the domains and they will travel the whole length of the sample. Clearly the transit time will then be about twice as great, and the frequency about half as great, as in the former case.

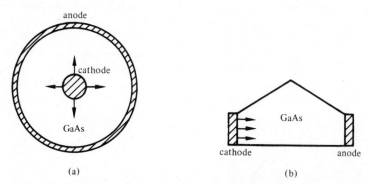

(a)                                         (b)

**Figure 1.18.** Devices based on the Gunn effect: (a) tunable oscillator; (b) voltage-controlled frequency switch.

### 1.5 Acoustic amplifiers
The amplification of sound waves by a semiconductor was first reported by Hutson *et al.* (1961). This amplification resulted from the interaction of the sound waves with charge carriers, which were drifting under the influence of an electric field with a velocity of the order of the velocity of sound. The necessarily strong interaction was achieved by using the piezoelectric semiconductor cadmium sulphide, CdS.

The importance of ultrasonic amplification is that, like the Gunn effect, it overcomes the limitations on frequency that are imposed on semi-conductor junction devices by their structure. There is, in fact, a good analogy between the ultrasonic amplifier and the travelling wave tube.

Ultrasonic amplification also provides the means of obtaining lossless delay lines.

A good physical explanation of the phenomenon has been given by Pippard (1963), who views it as an example of stimulated emission. Pippard considers the simple case of a semiconductor having spherical surfaces of constant energy centred at $\mathbf{k} = 0$ in the absence of current flow, and assumes a constant relaxation time. It is supposed that the product of phonon wavenumber $q$ (which is always much less than the electron wavenumber $k$) and electron free path length is much greater than unity, though this is not a necessary condition for amplification. This assumption implies that there must be strict conservation of wavevector as well as energy in the interaction. It can then be shown that phonons, moving along the $x$ axis with a velocity $v_s$, interact only with electrons which have a component of velocity in the $x$ direction equal to $v_s$. The surface of interaction is given by the plane $k_x = m^*v_s/\hbar$, where $m^*$ is the effective mass.

Figure 1.19a shows how a phonon with wavevector $\mathbf{q}$ interacts with an electron of wavevector $\mathbf{k}_1$, changing its wavevector to $\mathbf{k}_2$, when the current is zero. The process in the upper part of the diagram for which $k_1 < k_2$ represents phonon absorption (attenuation of the sound waves), whereas the process in the lower part of the diagram represents phonon emission. These processes are equally probable apart from a factor $f_1(1 - f_2)$, where

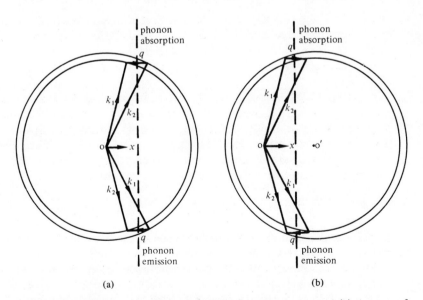

**Figure 1.19.** Electron-phonon interaction for (a) zero current and (b) a current for which the drift velocity of the electrons exceeds the velocity of sound. $\mathbf{k}_1$ is the initial electron wavevector, $\mathbf{k}_2$ the final electron wavevector, and $\mathbf{q}$ the phonon wavevector. The dashed line represents the plane of interaction.

$f_1$ and $f_2$ are the occupation numbers of the initial and final electron states. In the situation shown in Figure 1.19a, $f_1 > f_2$ for phonon absorption, while $f_2 > f_1$ for phonon emission. Thus, phonon absorption is more probable than phonon emission and the overall effect of the electron interaction is to attenuate the ultrasonic waves.

Now consider the semiconductor to carry such a large current that the drift velocity of electrons in the $x$ direction exceeds the velocity of sound. This may be regarded as a shift of the centre of the constant energy surfaces to a point O' in wavevector space, such that $OO' > m*v_s/\hbar$, as shown in Figure 1.19b. The plane of interaction now lies between O ($k = 0$) and O', so that, since the occupation number decreases with increase of energy, $f_1 < f_2$ for phonon absorption and $f_1 > f_2$ for phonon emission. In other words, when the drift velocity exceeds the sound velocity, the emission process is the more probable and there is sound wave amplification.

Hutson and White (1962) derived an expression for the attenuation constant $\alpha$ from the piezoelectric stress–strain relationships. The electric fields associated with the propagating sound waves can be related to the fields associated with the charge carriers and, thus, a complex elastic stiffness constant found. The attenuation constant is given by

$$\alpha = \frac{K^2\omega_r}{2\gamma v_s}\left[1+\frac{\omega_r^2}{\gamma^2\omega^2}\left(1+\frac{\omega^2}{\omega_r\omega_d}\right)^2\right]^{-1}, \tag{1.20}$$

where $K^2$ is the piezoelectric coupling constant, $\omega_r$ is the conductivity relaxation frequency given by $\sigma/\epsilon$, $\omega_d$ is the diffusion frequency given by $v_s^2/D_n$, $D_n$ is the diffusion coefficient for electrons, $\omega$ is the frequency of the sound waves, and $\gamma = 1-v_D/v_s$, $v_D$ being the drift velocity of the electrons.

Equation (1.20) shows us that the attenuation reaches a maximum when $\gamma = (\omega_r/\omega)(1+\omega^2/\omega_r\omega_d)$ and a minimum when $\gamma = -(\omega_r/\omega)(1+\omega^2/\omega_r\omega_d)$. A typical theoretical plot of attenuation against $\gamma$ is given in Figure 1.20. The attenuation (positive or negative) is strongest when the frequency is given by $\omega^2 = \omega_r\omega_d$; at much lower frequencies the maximum and minimum values of the attenuation constant are given by

$$\alpha_{\min}^{\max} = \pm\frac{K^2\omega}{4v_s}. \tag{1.21}$$

In practice the theoretical predictions may be modified by trapping effects, which are often very pronounced in CdS.

Figure 1.21 shows how ultrasonic amplification may be obtained in cadmium sulphide for the range of frequency 10-100 MHz. A single crystal of CdS is cut to a rectangular shape and the end faces are polished flat and parallel. Metal layers on these end faces serve as ohmic contacts and as bonds for the quartz transducers. The d.c. field which causes the drift of the carriers is pulsed so as to avoid excessive power dissipation.

The carriers themselves are produced in the required concentration by photo-excitation using light from a CdS filter.

Some typical experimental results are given in Figure 1.22 which shows the relative attenuation plotted against applied voltage for transverse sound waves at 45 MHz (Harcourt *et al.*, 1966). The zero of attenuation is taken to be the value for the crystal in the dark with no applied field. Overall gain has actually been reported for frequencies up to about 1 GHz.

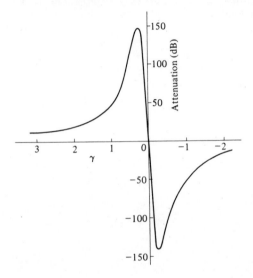

**Figure 1.20.** Typical plot of attenuation against $\gamma$ as predicted by Hutson and White's theory. $\gamma$ is equal to $1 - v_D/v_s$, where $v_D$ is the drift velocity of the electrons and $v_s$ is the velocity of sound.

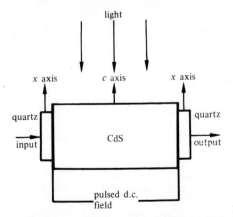

**Figure 1.21.** Experimental arrangement for observing ultrasonic amplification in cadmium sulphide.

It should be noted that the formation of high-field domains is possible as a result of the acousto-electric effect. These domains travel with the velocity of sound, which is two orders of magnitude smaller than the velocity of the Gunn domains. Although this means that microwave oscillators using acousto-electric domains are not feasible, it does allow the motion of domains to be studied more easily, and there remains the possibility of many practical devices based on the principle of current modulation through local changes of the product of area and carrier concentration (see end of Section 1.4).

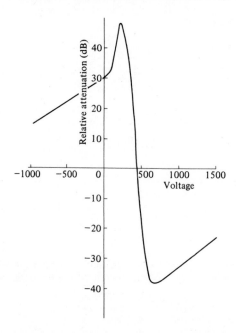

**Figure 1.22.** Typical experimental results.

**References**
Altenkirch, E., 1909, *Z. Physik.*, **10**, 560.
Altenkirch, E., 1911, *Z. Physik.*, **12**, 920.
Austin, I. G., Springthorpe, A. J., Smith, B. A., and Turner, C. E., 1967, *Proc. Phys. Soc. (London)*, **90**, 157.
Blakemore, J. S., Schultz, J. W., and Myers, J. G., 1962, *Rev. Sci. Instr.*, **33**, 545.
Bott, I. B., 1968, *Phys. Bull.*, **19**, 144.
Butcher, P. N., 1967, *Rept. Progr. Phys.*, **30**, 97.
Carroll, J. A., and Spivak, J. F., 1966, *Solid-State Electron.*, **9**, 383.
Christian, J. W., Jan, J. P., Pearson, W. B., and Templeton, I. M., 1958, *Can. J. Phys.*, **36**, 627.
Cope, R. G., and Penn, A. G., 1968, *Brit. J. Appl. Phys.*, **1**, 161.
Futaki, H., 1965, *Japan. J. Appl. Phys.*, **4**, 28.
Goldsmid, H. J., 1960, **Applications of Thermoelectricity** (Methuen, London).

Gunn, J. B., 1963, *Solid-State Commun.*, **1**, 88.
Harcourt, R. W., Froom, J., and Sandbank, C. P., 1966, *Radio Electron. Engr.*, **31**, 145.
Heikes, R. R., and Johnston, D. M., 1957, *J. Chem. Phys.*, **26**, 582.
Hilsum, C., 1961, *Brit. J. Appl. Phys.*, **12**, 85.
Hutson, A. R., McFee, J. H., and White, D. L., 1961, *Phys. Rev. Letters*, **7**, 237.
Hutson, A. R., and White, D. L., 1962, *J. Appl. Phys.*, **33**, 40.
Ioffe, A. F., 1957, *Semiconductor Thermoelements and Thermoelectric Cooling* (Infosearch, London).
Morin, F. J., 1959, *Phys. Rev. Letters*, **3**, 34.
Pippard, A. B., 1963, *Phil. Mag.*, **8**, 161.
Putley, E. H., 1960, *The Hall Effect and Related Phenomena* (Butterworths, London).
Ross, I. M., and Saker, E. W., 1957, *Nature*, **179**, 146.
Ross, I. M., and Thompson, N. A. C., 1955, *Nature*, **175**, 518.
Scarr, R. W. A., and Setterington, R. A., 1960, *Proc. Inst. Elec. Engrs. (London), Pt.B*, **107**, 395.
Weiss, H., 1966, *Solid-State Electron.*, **9**, 443.
Whitlock, W. H., and Hilsum, C., 1960, *Nature*, **185**, 302.
Wieder, H. H., 1966, *Solid-State Electron.*, **9**, 373.

# Semiconductor diodes

The wide range of phenomena that occur in single p-n junctions result in devices whose applications cover very diverse fields. A few of these devices and applications will be described here.

## 2.1 The low-frequency rectifier

If a sample of n-type semiconducting material were brought into contact with a p-type sample, in such a way that continuity of the crystal lattice existed across the junction, then the higher concentration of electrons in the n-type material would result in electrons crossing from the n- to the p-type material, and, in a similar way, holes would cross from the p- to the n-type semiconductor. Both charge movements would result in the n-type material becoming positively charged with respect to the p-type, and a potential difference would build up across the junction.

The polarity of such a potential difference is clearly such as to oppose the charge flow that is responsible for it, so that eventually the system reaches equilibrium with no net movement of charge across the junction.

In equilibrium, carrier generation is balanced by carrier recombination, so that the total number of charge carriers remains constant within any region of the device. Under these conditions, Maxwell's energy equation may be used to determine the contact potential difference across the junction. At distances well removed from the disturbance of the junction, the majority carrier densities $n_n$ and $p_p$ will simply be equal to the respective donor and acceptor densities $N_D$ and $N_A$ on the n-type and p-type sides. Minority carrier densities will be denoted as $n_p$ and $p_n$ (electrons in the p region, and holes in the n region, respectively). The difference in energy between electrons on the p side of the junction and those on the n side, is just $eV_c$, where $V_c$ is the contact potential difference. Hence, applying Maxwell's equation,

$$n_p \propto n_n \exp\left(\frac{eV_c}{kT}\right)$$

and

$$p_n \propto p_p \exp\left(\frac{eV_c}{kT}\right)$$

(the constant of proportionality being the same in both cases) we get

$$V_c = \frac{kT}{e}\ln\left(\frac{n_p}{n_n}\right) = \frac{kT}{e}\ln\left(\frac{p_n}{p_p}\right) = \frac{kT}{e}\ln\left(\frac{n_i^2}{N_A N_D}\right), \qquad (2.1)$$

where $n_i$ is the intrinsic carrier concentration. Since $e$ is positive, and

and $N_A N_D > n_i^2$, $V_c$ is negative and

$$|V_c| = \frac{kT}{e} \ln \frac{N_A N_D}{n_i^2} \quad .$$

Typical values of $|V_c|$ are around 600-700 mV for silicon junctions and 200-300 mV for germanium. The difference in values reflects the difference between the intrinsic carrier concentrations in silicon and germanium. It is possible to quote typical values because $V_c$ is not directly proportional to the doping densities. For example, a tenfold change in both $N_A$ and $N_D$ will alter the $V_c$ by not much more than 100 mV.

Equation (2.1) is perfectly general and may be used to relate the charge density at any point $x$ to the potential at $x$:

$$V(x) = \frac{kT}{e} \ln \left[ \frac{n(x)}{n_n} \right]. \tag{2.2}$$

A complete solution for charge and potential distributions as functions of $x$ may then be derived as follows.

Since recombination is assumed negligible, and the overall junction current is zero, both the net hole current and the net electron current must each individually be zero ($j_n = j_p = 0$). Therefore electron drift must exactly balance electron diffusion, that is

$$\mu_n n(x) \mathcal{E}(x) = -D_n \frac{dn(x)}{dx} \tag{2.3a}$$

and similarly for holes

$$\mu_p p(x) \mathcal{E}(x) = D_p \frac{dp(x)}{dx} \quad , \tag{2.3b}$$

where the mobility $\mu$ and diffusion coefficient $D$ are related by Einstein's equation,

$$\frac{\mu}{D} = \frac{e}{kT} \quad .$$

Furthermore,

$$\epsilon \frac{d\mathcal{E}(x)}{dx} = \rho(x) \quad \text{(Gauss's law)}$$

where the charge density

$$\rho(x) = [N_D - N_A + p(x) - n(x)]e$$

and

$$\mathcal{E}(x) = -\frac{dV(x)}{dx} \quad .$$

A logarithmic plot of typical charge carrier densities is shown in Figure 2.1a. As Figure 2.1b is plotted on a linear scale, it can be seen that there is a well-defined region, AB, virtually devoid of charge carriers. This is the depletion layer and its properties are crucial to the behaviour of the p-n junction.

As there are comparatively few charge carriers in the depletion layer, there will be a net charge density due to the ionised impurity atoms firmly bound in the lattice. The charge density will therefore be $+N_D e$ in the n-type material and $-N_A e$ in the p-type.

The rectangular charge profile, shown in Figure 2.2a, gives rise to a triangular field profile, Figure 2.2b, and an overall voltage profile as shown in Figure 2.2c. The peak field in the depletion layer, $\hat{\mathscr{E}}$, is given by

$$-\hat{\mathscr{E}} = \frac{N_D l_D e}{\epsilon} = \frac{N_A l_A e}{\epsilon}$$

and, by integration, the contact potential difference is given by

$$V_c = -\left(\frac{N_A l_A^2 e}{2\epsilon} + \frac{N_D l_D^2 e}{2\epsilon}\right). \tag{2.4a}$$

Also,

$$V_c = \frac{kT}{e} \ln\left(\frac{n_i^2}{N_A N_D}\right). \tag{2.4b}$$

These equations show the qualitative effects of varying the doping levels. Increasing the doping densities results in (a) an increase in contact potential difference $V_c$, (b) a reduction in depletion layer width, and (c) an increase in peak depletion layer field strength.

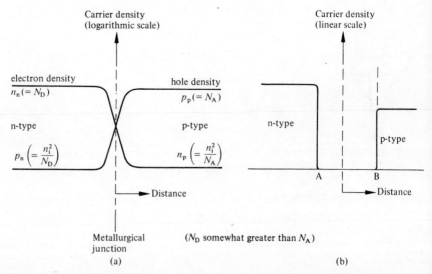

**Figure 2.1.** Carrier densities near a p-n junction: (a) logarithmic scale; (b) linear scale.

In equilibrium, there is no overall flow across the depletion layer and, also, the carrier densities are very low there. However, it would be incorrect to assume that there is no significant movement of carriers across the depletion layer.

Although the total currents are zero, individual hole and electron currents may be found by evaluating the separate terms in Equation (2.3). A rough estimate may be made by assuming a linear fall in carrier density across the junction. If we assume, for example, a germanium junction with $N_A = N_D = 10^{23}$ m$^{-3}$, the depletion layer width from Equations (2.4) is about $5 \times 10^{-8}$ m. The electron diffusion current is then very approximately given by

$$j_n \text{ (diff)} = -D_n e \frac{dn(x)}{dx} \approx 3 \times 10^3 \text{ A mm}^{-2} \text{ of junction area.}$$

This is an extremely high current density, and is some orders of magnitude greater than the net current densities due to normal external currents. The point is very important as it means that when the diode is subject to external bias, thermodynamic equilibrium is hardly disturbed and Maxwell's equation can still be applied.

The large values of the individual charge movements also show that charge carriers have little difficulty in crossing the depletion layer, either way, almost irrespective of the high field that is present. If in some way the concentration of carriers on either side of the junction is raised above the equilibrium value, there will be an increase in the flow of those carriers to the other side of the junction, and a fall in the flow in the opposite direction. This happens virtually independently of the charge carrier polarity with respect to the depletion layer field.

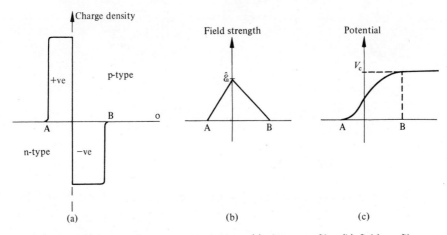

**Figure 2.2.** Depletion layer at a p–n junction: (a) charge profile; (b) field profile; (c) voltage profile.

A qualitative view of some of the above features is provided by the energy band approach (Blakemore *et al.*, 1953). Doping a semiconductor moves the Fermi level away from its (near) midband position. Energy band diagrams appropriate to the separate p and n material are shown in Figure 2.3a. On contact, holes and electrons diffuse across the junction (Figure 2.3b). This results in the p-type material becoming negative with respect to the n-type and the energy levels on the p side are raised with respect to those on the n side. Eventual equilibrium is reached when the Fermi level is constant throughout as in Figure 2.3c.

Figure 2.3c also shows that majority carriers of either polarity may readily cross the depletion layer provided that their thermal energies are greater than the energy $eV_c$. This will be so, for example, for electrons with energy $Z$, or holes with energy $Z_1$. Such carriers will, even in 'equilibrium', be continually crossing and re-crossing the depletion layer in a random Brownianlike motion. Clearly an increase in carrier density on one side of the junction will lead to a net flow of those carriers across the junction.

The common statement that depletion layer fields accelerate carriers across to the other side (if we assume correct polarity, of course) is rather a simplified explanation of what occurs.

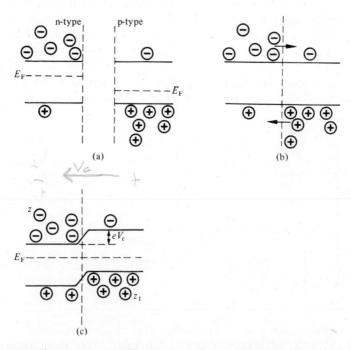

**Figure 2.3.** Energy diagram for a p–n junction (a) p and n regions isolated, (b) regions brought into contact, and (c) equilibrium reached.

The low carrier densities in the depletion layer result in a much lower conductivity than in the bulk semiconductor. Application of an external bias $V_{app}$ will cause an almost equal change in voltage across the depletion layer, with only a small voltage drop occurring in the undepleted, bulk semiconductor.

Within the bulk semiconductor, the small field will give rise to carrier drift in the appropriate direction. This will be a majority carrier effect, since minority carriers are present only in very small numbers. The drift of the majority carriers will give rise to a corresponding net flow across the depletion layer. This results in a change in the minority carrier density on the other side of the junction. Continuity of charge movement therefore implies minority carrier movement away from the junction, on the other side. However, as there is insufficient field to produce minority carrier drift, continuity is maintained by minority carrier diffusion.

Under external bias, holes will drift/cross/diffuse from p- to n-type, while electrons will drift/cross/diffuse in the opposite direction. It is interesting to note that in the p-type semiconductor, for example, majority carrier drift (holes) and minority carrier diffusion (electrons) go on simultaneously but quite independently of each other.

Actual current densities may be found by considering the minority carrier diffusion currents away from the depletion layer. In the absence of recombination, these currents are proportional to $dn_p/dx$ and $dp_n/dx$, so that current continuity implies constant minority carrier gradients.

At the ohmic contacts to the external circuit, minority carrier densities are maintained at the levels $n_p$ and $p_n$, whereas at the edges of the depletion layer they will have different values, denoted here by $n'_p$ and $p'_n$ respectively. In the p material, then, there is a minority carrier gradient $dn_p/dx = (n'_p - n_p)/L_p$ and in the n material $dp_n/dx = (p'_n - p_n)/L_n$ as shown in Figure 2.4a. The minority carrier charge above the unbiased value is known as the excess charge and is given by

$$Q_s = e \sum (p'_n - p_n) + e \sum (n'_p - n_p) ;$$

it is shown shaded in the relevant diagrams. Clearly, the total current $(I_n + I_p)$ will be proportional to $Q_s$ in a given diode.

In practice, the region of the semiconductor away from the depletion layer is extensive so far as carriers are concerned, being many diffusion lengths in size. Under these circumstances, recombination is not normally negligible. Since recombination is proportional to the excess charge density at every point, there will now be an exponential decrease in minority density from the depletion layer to the external contacts, as shown in Figure 2.4b.

Although it is no longer true to say that the current $I_n$ equals $eD_n(n'_p - n_p)/L_p$ it is still nevertheless proportional to this quantity, i.e.

$$I_n \propto (n'_p - n_p) \,,$$

and similarly

$$I_p \propto (p'_n - p_n) \,.$$

(Also the total current will still be proportional to $Q_s$.)

The carrier densities $p'_n$ and $n'_p$ at the edges of the depletion layer may be found by applying Maxwell's equation again. The internal junction potential is now $V_{app} + V_c$ rather than just $V_c$, so that

$$n'_p = n_n \exp\left[\frac{e(V_{app} + V_c)}{kT}\right]$$

and

$$p'_n = p_p \exp\left[\frac{e(V_{app} + V_c)}{kT}\right].$$

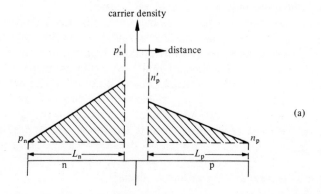

(a)

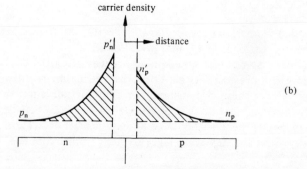

(b)

**Figure 2.4.** Minority carrier densities near a p-n junction (a) without recombination and (b) with recombination.

Therefore, an electron current $I_n$ flows, given by

$$I_n \propto -n_n \left\{ \exp\left[\frac{e(V_{app} - |V_c|)}{kT}\right] - \exp\left(-\frac{|V_c|}{kT}\right) \right\} \qquad (2.5)$$

and, similarly, a hole current $I_p$ given by

$$I_p \propto -p_p \left\{ \exp\left[\frac{e(V_{app} - |V_c|)}{kT}\right] - \exp\left(-\frac{|V_c|}{kT}\right) \right\}. \qquad (2.6)$$

Thus the total current $I$, equal to $I_n + I_p$, is given by

$$I = I_n + I_p \propto \left[\exp\left(\frac{eV_{app}}{kT}\right) - 1\right].$$

Since $kT/e$ is about 25 mV, the exponential will greatly exceed unity if $V_{app}$ considerably exceeds 25 mV. Thence,

$$I \propto \exp\left(\frac{eV_{app}}{kT}\right).$$

With positive values of $V_{app}$, $I$ increases rapidly with $V_{app}$. The junction is said to be forward-biased. With negative values of $V_{app}$, the device is reverse-biased. The total current is still given by the expression

$$I \propto \left[\exp\left(\frac{eV_{app}}{kT}\right) - 1\right]$$

but now, if $V_{app}$ is significantly more negative than $-25$ mV, the exponential will become negligible with respect to unity. In other words, the reverse current should saturate. A typical diode characteristic is shown in Figure 2.5.

In practice, a saturated value of the reverse current may not always be observed. The effects of carrier recombination and generation within the

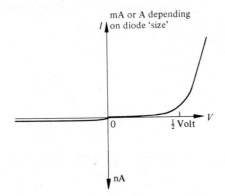

**Figure 2.5.** Typical silicon diode characteristic.

depletion layer have so far been neglected. When forward-biased, the product of hole and electron densities within the depletion layer is increased above the intrinsic value $n_i^2$, and there must be a fall in current due to net carrier recombination. The overall change in current clearly depends on the volume, or width, of the depletion layer, and in forward conduction this is sufficiently small for the resultant effect on the forward current to be negligible. When the junction is reverse-biased, the carrier densities are reduced below the intrinsic values and generation now outweighs recombination. However, the width of the depletion layer increases in reverse conduction, and the resultant increase in reverse current due to net carrier generation may be sufficient to mask the basic saturation characteristic. The variation in width with reverse bias is such that this additional contribution to the reverse current will vary approximately as the square root of the applied voltage.

It can also be seen from Equations (2.5) and (2.6) above, that, for a given value of $V_{app}$, $I_n$ may be made very much greater than $I_p$ by making $n_n \gg p_p$; in other words, the hole current may be made negligible in comparison with the electron current by doping the n-type material more heavily than the p-type. Under these conditions the n-type semiconductor could be called the emitter, and the p-type the collector. Similarly the hole current may be made predominant by doping the p-type material more heavily.

It should be stressed that the only direct effect of $V_{app}$ is to change the minority densities at the edge of the depletion layer. Any current flow is an indirect effect and its magnitude and even direction may well be controlled by conditions remote from the junction, almost totally independent of $V_{app}$ (see Chapter 3).

Qualitatively the same general characteristics emerge from a look at the energy band diagrams, Figure 2.6. Forward biasing raises the n-type levels with respect to the p-type levels so that a net flow of carriers can cross the junction (Figure 2.6a). Taking into account the higher densities

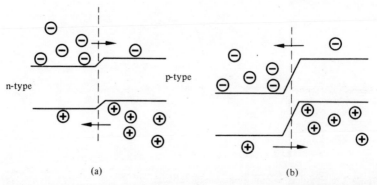

(a)                                                    (b)

**Figure 2.6.** Energy diagram for (a) forward-biased and (b) reverse-biased junctions.

that can take part as the external bias is steadily increased, we can see that the rate of rise in current with voltage will be more than linear.

Similarly, reverse biasing the junction soon leads to the situation where only a few minority carriers can take part in conduction, and the current flow will not increase with further reverse bias, as in Figure 2.6b.

## 2.2 Junction breakdown and Zener diodes

If sufficient reverse bias is applied to a junction, the field strength across the depletion layer will become large enough for the junction to break down. At this stage the junction current will increase rapidly and, unless the increase is limited, the junction will be damaged.

There are two mechanisms that result in diode breakdown. Which of the mechanisms is dominant depends on the doping density in the semiconductor.

Avalanche breakdown (Wolff, 1954) occurs if carriers accelerating within the depletion layer pick up sufficient energy to ionise lattice atoms. Each ionisation event produces a pair of charge carriers which will be separated by the depletion layer field and, if those carriers can pick up sufficient energy to produce further lattice ionisation, avalanche breakdown will occur.

Avalanche breakdown is possible only with reasonably wide depletion layers; otherwise, the carriers will not have a chance to produce ionisation and further multiplication. This implies a lightly doped semiconductor. Because of the light doping, $\hat{\mathscr{E}}$ is initially small, and comparatively large values of $V_{app}$ are necessary before avalanche breakdown can happen (McKay and McAfee, 1953).

With heavier doping, the depletion layer width decreases until it is comparable with the mean free path of the carriers. Despite the high values of $\hat{\mathscr{E}}$, avalanche breakdown cannot occur. Under these conditions, breakdown comes about by electron tunnelling (Zener, 1934). The high field strength and narrow depletion layer combine to give a very real probability of electrons tunnelling straight through the forbidden energy gap separating the n-type conduction band on one side of the depletion layer from the p-type valence band on the other. Since $\hat{\mathscr{E}}$ is already large, tunnelling breakdown will be predominant in junctions having low breakdown voltages.

Energy band diagrams, for reverse-biased junctions, which help to explain the two types of breakdown are shown in Figure 2.7.

Diodes intended for voltage stabilisation by operating on the breakdown part of the characteristics are called Zener diodes, although, strictly speaking Zener breakdown specifically implies the tunnelling effect. The crossover between avalanche and tunnelling is found for a reverse breakdown voltage $V_{app}$ of about $5 \cdot 3$ V. Zener diodes operating at this voltage have the advantage of being comparatively temperature independent since the two breakdown effects have opposite temperature coefficients, which, in fact, just balance out at the crossover voltage.

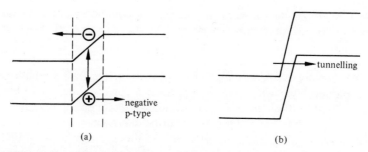

(a)            (b)

**Figure 2.7.** Junction breakdown mechanisms: (a) avalanche breakdown; (b) tunnelling.

## 2.3 Conventional diodes at high frequencies

Although the 'ordinary' p-n junction is an effective rectifier at low frequencies, its performance falls off at higher frequencies. Figure 2.4b shows the minority charge distribution of a forward-biased diode. If the applied voltage is suddenly reversed, then, as seen earlier, the direct effect is to depress the minority carrier concentration at the edge of the depletion layer. The charge distribution of the diode becomes as shown in Figure 2.8, and the very steep gradients at $X$ and $Y$ mean that a considerable reverse current will flow. The reverse current will continue to flow until the excess minority carriers either have crossed the depletion layer, or have recombined with oppositely charged carriers (Henderson and Tillman, 1957).

In effect then, a forward-conducting diode builds up an excess minority charge concentration $Q_s$ such that, on reversing the applied voltage, a reverse current $I_r$ can flow for time $\Delta t$, where $Q_s = \int_0^{\Delta t} I_r \, dt$. During this time the diode is not acting as a rectifier, since it is able to pass a reverse current. A similar situation arises when a reverse-biased diode is suddenly forward-biased; there is a time lag before the appropriate minority charge density builds up so that the expected forward current can pass. In

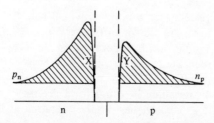

**Figure 2.8.** Minority carriers densities at switch-off.

resistive circuits, the reverse current is often limited to values similar to
that of the forward current $I_f$ and the time lag $\Delta t$ is roughly equal to
$Q_s/I_f$.

Now it is the distribution of $Q_s$ within the semiconductor that results in
diffusion current flow, that is

$$I_n = eD_n \frac{dn(x)}{dx} ,$$

where

$$Q_{s(n)} = \int \{n(x) - n_p\} e\, dx ,$$

with a similar expression for $I_p$. Thus $Q_s$ is proportional to $I$, and,
therefore, the delay in response is a constant for a given diode. In other
words, diodes have a natural cut-off frequency above which they are not
suitable for normal rectification purposes.

The presence of the excess charge in a conducting diode gives the diode
a capacitive characteristic. Any change in applied voltage results in a
change in diode current and, hence, a change in the stored charge. It is
referred to as the diffusion capacitance.

Attempts to speed up the response of the p-n junction diode are usually
aimed at reducing the diffusion capacitance, by reducing the stored charge
associated with a given forward current. This can be achieved by
deliberately lowering the minority carrier lifetime or by cutting the
thickness of semiconductor normal to the plane of the actual junction, or
both.

Figure 2.4 shows minority charge distributions in diodes with negligible
and not quite negligible recombination. If the semiconductors are doped
additionally with impurities such as gold, carrier recombination increases,
lifetimes decrease, and the charge distribution is modified to that of
Figure 2.9. Gold-doped diodes are readily available, and the reduced
charge storage leads to cut-off frequencies of up to hundreds of megahertz.

If one side of a semiconductor is more heavily doped than the other,
that side acts as an emitter, and the total stored charge is almost entirely

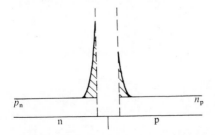

**Figure 2.9.** Minority carrier contribution with high recombination.

that within the collector (see Figure 2.10). For the same applied voltage, reduction of the collector width must alter the minority distribution as shown in Figure 2.11. Not only is the stored charge reduced, but at the same time the gradient at X is increased; in other words, a higher forward current is involved. Nevertheless, diodes whose switch-off time depends on minority carrier effects are hardly ever likely to be effective rectifiers in the gigahertz range.

However, the very reason for failure in high-frequency rectification allows the diode to be used as a very effective on–off switch up to many gigahertz (Uhlir, 1958b).

If an 'on' diode is subject to an alternating signal voltage, whose frequency is well above the diode cut-off, then the diode will not be able to go 'off' during a reverse half-cycle before the next 'on' half-cycle comes along. The reduction in $Q_s$ during the reverse current flow will then be restored. In effect the diode is conducting in either direction the whole time.

If the reverse charge flow exactly equalled the forward flow, $Q_s$ would, nevertheless, gradually diminish because of carrier recombination, and the diode would then go 'off'. To keep the diode 'on' a small direct current is superimposed on the r.f. current (Figure 2.12). The diode may then be turned 'off' at will by switching off the small d.c. bias.

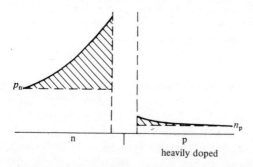

**Figure 2.10.** Diode with unequal doping on either side of the junction.

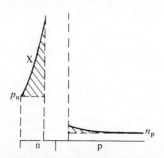

**Figure 2.11.** Diode with narrow collector.

A few numerical values might help to demonstrate the efficiency of the system. A certain diode passing a net forward current of 100 mA has a stored charge $Q_s$ of $10^{-8}$ C. (It may be noted that this corresponds to a cut-off frequency, $I/Q_s$, of about 10 MHz.) At 1 GHz this charge will be dissipated in a reverse half-cycle by a current of about 20 A. In other words, a 20 A current, alternating at 1 GHz, may be superimposed on the 100 mA net forward current. Suppose the 20 A r.f. signal passes along a 70 Ω transmission system. This means a power flow of about 3 kW. The d.c. power needed to bias the diode at 100 mA, and at an average forward voltage of perhaps 0·5 V, is about 0·05 W, which is a factor of $6 \times 10^4$ below the power that is being switched.

When the d.c. bias is disconnected, $Q_s$ will gradually disperse, during a time inversely proportional to the diode cut-off frequency. Although it may be several hundreds of r.f. cycles before r.f. transmission is completely blocked, this will, in fact, only take a small fraction of a second.

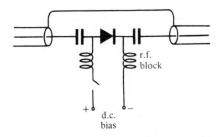

**Figure 2.12.** Diode used as a microwave switch.

### 2.4 Backward diodes

It has been shown that increasing the doping densities raises the depletion layer field strength, and reduces the depletion layer width. Both of these factors encourage tunnelling of electrons from n-type material to p-type when the diode is reverse-biased, and also lower the reverse bias necessary to break the junction down, i.e. the bias at which tunnelling current suddenly increases with reverse voltage at a rapid rate.

If the doping densities are sufficiently high, $\hat{\mathscr{E}}$ will be large enough to break down the diode even before any reverse bias is applied. The diode will then behave as a comparatively good conductor in the reverse direction, but as a normal diode in the forward direction (Burrus, 1963). The overall characteristic will be as shown in Figure 2.13.

The effect on the Fermi level of doping a semiconductor has been referred to earlier (see Figure 2.3a, for example). If a diode is just in the breakdown regime with zero applied bias, it means that the Fermi levels have been shifted up to the bottom of the n-type conduction band, and down to the top of the p-type valence band (see Figure 2.14a).

On applying a reverse bias, the energy diagram changes to that of

Figure 2.14b, and a tunnelling current (breakdown) flows in addition to, and greatly exceeding, the usual reverse current. In the forward direction (Figure 2.14c), conduction is entirely as usual, so that the whole characteristic is as in Figure 2.13.

So far as small-signal voltages are concerned, the diode appears to have a backward characteristic; conduction occurs much more readily in the reverse direction than in the forward direction (hence the name 'backward diode'). It will also be observed that an ordinary diode would show hardly

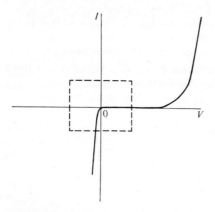

**Figure 2.13.** Characteristic of a backward diode (see text for discussion of region enclosed by dashed line).

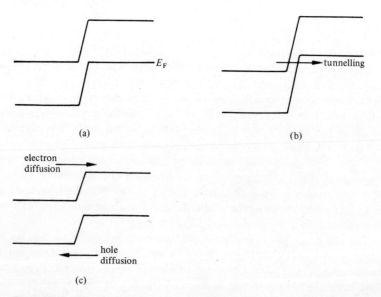

**Figure 2.14.** Energy diagram of a backward diode: (a) zero bias; (b) reverse bias; (c) forward bias.

any conduction characteristics within the range of applied voltage bounded by the dashed line in Figure 2.13. Clearly then, the backward diode finds useful application in the rectification or detection of very small signals for which the orthodox diode would be quite useless.

A further feature of the backward diode of great practical value is its very rapid response to changes in the applied voltage. The frequency response of any device is related to the time taken for equilibrium to be regained after a change in external conditions. In the backward diode, the tunnelling currents are entirely majority carrier effects; the minority carriers are quite undisturbed. The response time of the diode is, therefore, closely related to the majority carrier relaxation time, which in practice is normally well below a nanosecond. Backward diodes may be used at frequencies of up to tens of gigahertz, their low-voltage characteristic making them particularly suitable for use in, for example, microwave receivers.

### 2.5 Tunnel diodes

By increasing the semiconductor doping, the reverse breakdown voltage of the ordinary diode may be reduced to zero as in the backward diode. By still further increasing the doping, diodes can be formed which are not only broken down with zero applied bias, but which, in fact, require a forward voltage (opposing $V_c$) to prevent depletion layer breakdown. These are tunnel diodes. The very heavy doping results in the Fermi energy level being raised above the bottom of the n-type conduction band, and below the top of the p-type valence band. The energy diagram of an unbiased tunnel diode is shown in Figure 2.15a.

On forward biasing, as in Figure 2.15b, a small 'normal' forward current will flow, but the main conduction occurs by tunnelling from the higher electron density at the bottom of the n-type conduction band, to the lower electron density at the top of the p-type valence band. At first the tunnelling current will increase with $V_{app}$, but eventually the top of the p-type valence band falls below the bottom of the n-type conduction band, as in Figure 2.15c, and tunnelling is no longer possible. This normally happens when $V_{app}$ is of the order of 200–300 mV. Further increase in $V_{app}$ results in the 'normal' forward-biased junction current. The overall characteristic is thus the sum of the normal diode characteristic and that due to tunnelling (Esaki, 1958). A typical example is given schematically in Figure 2.16.

The negative-resistance region may be used for the generation of sinusoidal waveforms. Oscillations induced in a simple LC circuit, as shown in Figure 2.17a, will gradually die away owing to unavoidable losses ($R_L$), unless energy is supplied to make up those losses. This can be achieved with a tunnel diode circuit, Figure 2.17b, provided that the negative resistance of the tunnel diode is sufficient to outweigh the positive loss resistance.

The tunnelling is a majority carrier phenomenon so that, as with the backward diode, the response time is very short. Tunnel diode oscillators can readily be built with operating frequencies of some gigahertz (Schneider, 1963). The amplitude of the oscillation is limited by the low

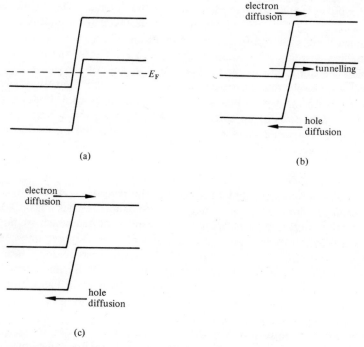

**Figure 2.15.** Energy diagram of a tunnel diode: (a) zero bias; (b) forward bias; (c) more forward bias.

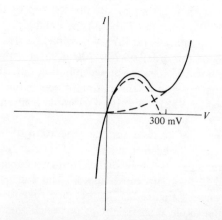

**Figure 2.16.** Tunnel diode characteristic.

voltage that is characteristic of the negative-resistance region, but the tunnel diode oscillator is attractive because of the very high frequencies of oscillation, and the simplicity of the circuitry.

The tunnel diode is also very useful in high-speed pulse circuitry for generating rectangular pulses having sub-nanosecond edges. Consider a tunnel diode wired in series with a resistor $R_1$ across a comparatively slowly varying voltage $V_{app}$, as in Figure 2.18. As $V_{app}$ rises from zero, the operating point moves up the tunnel diode characteristic (region A of Figure 2.19). At a critical value of $V_{app}$, the operating point must jump discontinuously from $X_1$ to $X_2$. The operating point then continues to move up the characteristic (region B). On reducing $V_{app}$, the operating

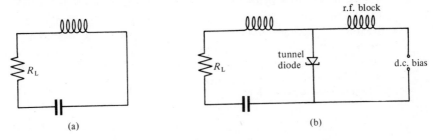

Figure 2.17. (a) Simple LC circuit. (b) Basic tunnel diode oscillator.

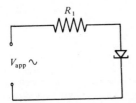

Figure 2.18. A tunnel diode switching circuit.

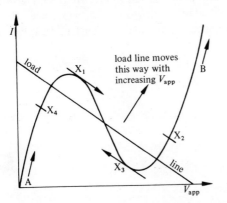

Figure 2.19. Characteristic of tunnel diode switch.

point moves steadily to $X_3$ and then again makes a discontinuous jump, to $X_4$, and finally moves steadily towards the origin.

A comparatively slowly varying voltage leads to a current waveform having very rapid discontinuities. Typically the change in current occurs in much less than a nanosecond, and the tunnel diode finds useful application as a switch in high-speed digital circuitry.

### 2.6 Varactors

The variation in width of the p-n junction depletion layer, as the applied voltage varies, implies a change in the associated charge profile. The junction therefore has a capacitive behaviour due to depletion layer effects, in addition to the diffusion capacitance mentioned earlier. It is interesting to note that the depletion layer capacitance is determined by the junction voltage, whereas the diffusion capacitance is determined by the junction current. This means, for example, that, under reverse bias, depletion capacitance is dominant, whereas in forward conduction the diffusion capacitance is generally dominant.

It has been shown that

$$|V_c| - V_{app} = \frac{N_A l_A^2 e}{2\epsilon} + \frac{N_D l_D^2 e}{2\epsilon} \ ,$$

where the bound charge $Q = -N_A l_A e = N_D l_D e$. Therefore, since the depletion capacitance $C_{dep}$ is $dQ/dV_{app}$,

$$C_{dep} = \left[ \frac{N_A N_D \epsilon e}{2(N_A + N_D)(|V_c| - V_{app})} \right]^{\frac{1}{2}} \ .$$

A typical variation of $C_{dep}$ with applied voltage is shown in Figure 2.20. It could refer to a germanium diode of junction area 1 mm² having $N_A = N_D = 10^{21}$ m⁻³.

The control of capacitance by an applied voltage finds many useful applications. An elementary one is in the automatic frequency control of radio receivers; a more sophisticated use involves the varactor as the lumped capacitance of a delay line. Parametric amplification relies on the

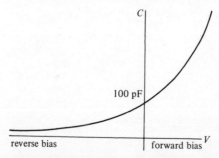

**Figure 2.20.** Variation of junction capacitance with bias.

non-linear characteristic of a loss-free device, and the reverse-biased varactor diode meets this requirement well at high frequencies (Manley and Rowe, 1956, 1958).

In many applications it is desirable that the variation of capacitance with applied voltage should be as great as possible. For an abrupt junction, $C$ is proportional to $1/V_{\text{app}}^{\frac{1}{2}}$, but a more favourable relation can result from different doping profiles across the junction.

A uniformly graded junction as in Figure 2.21a, results in triangular charge 'blocks' (Figure 2.21b) which means that $C$ varies as $1/V_{\text{app}}^{\frac{1}{3}}$. This is a slower capacitance variation with voltage, although, in practice, other features of the diode make it more useful than the abrupt junction mentioned above (Uhlir, 1958a).

Attempts to produce a more rapid variation of capacitance with applied voltage usually involve hyper-abrupt junctions as in Figure 2.22 (Chang

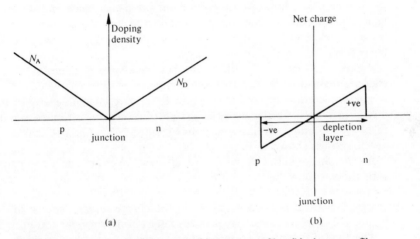

(a)                                                          (b)

**Figure 2.21.** Linear graded junction: (a) doping profile; (b) charge profile.

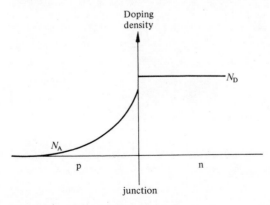

**Figure 2.22.** Hyper-abrupt junction.

*et al.*, 1963), and linear relations between $C$ and $1/V_{app}$ have been reported.

[The p-i-n diode, of Section 2.7, may also be looked on as a varactor diode with a 'two-part' $(C, V)$ characteristic.]

### 2.7 The addition of an intrinsic layer

Intrinsic semiconducting layers introduce further versatility to the p-n junction. Placed between the p- and n-type semiconductors, an intrinsic (i) layer will profoundly affect the depletion layer, whereas, sandwiched between slices of similar semiconductor, it will introduce a phase delay in current response due to the finite drift velocity of the carriers crossing it. Both possibilities lead to valuable practical devices.

If one imagined a p-i-n junction formed by bringing the three semi-conducting layers into simultaneous crystal-continuity contact, then, as in the p-n junction, holes would cross from the p-type material into the i region, and electrons from the n-type material into the same region. Similarly, electrons would cross from the i region to the p region, and holes from the i region to the n region.

Again, a full analysis would show that the charge densities in the i region are small compared with the 'exposed' positive charge density in the n-type semiconductor, and the 'exposed' negative charge density in the p-type material. Carrier densities are plotted on a logarithmic scale in Figure 2.23a and on a linear scale in Figure 2.23b. The net charge distribution is shown in Figure 2.24a.

As the charge densities within the intrinsic layer are small compared with those within the p and n regions, the whole width of the intrinsic region acts as a very wide depletion layer. Further, because of the small charge density, compared with that in the charge 'blocks' shown in Figure 2.24a, the intrinsic layer is virtually a uniform field region.

The field distribution corresponding to the Figure 2.24a is shown in Figure 2.24b, where

$$\mathcal{E}_i = -\frac{N_D l_D e}{\epsilon} = -\frac{N_A l_A e}{\epsilon} \ .$$

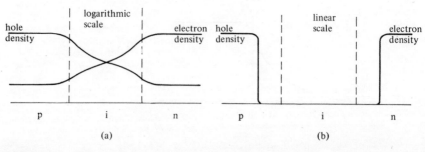

**Figure 2.23.** Carrier densities in p-i-n diode: (a) logarithmic scale; (b) linear scale.

In practice the width of the intrinsic layer is much greater than $l_D$ and $l_A$, so that the contact potential difference (see Figure 2.24c) is given by

$$V_c \approx \mathcal{E}_i w_i \, ,$$

$w_i$ being the thickness of the intrinsic layer. However, as before, $V_c$ may still be found from Maxwell's equation, namely

$$V_c = \frac{kT}{e} \ln\left( \frac{n_i^2}{N_A N_D} \right) ,$$

and is the same as it would have been without the intrinsic layer. Comparing the expressions for $V_c$ and $\mathcal{E}$ relevant to the p-n junction, with those above, setting $w_i \gg l_D$ and $l_A$, we can see that $\mathcal{E}_i$ must be very much less than the peak depletion-layer field in the basic p-n junction. This further implies that the bound charge blocks $eN_A l_A$, $eN_D l_D$ in Figure 2.24a, must be smaller than those of the p-n junction.

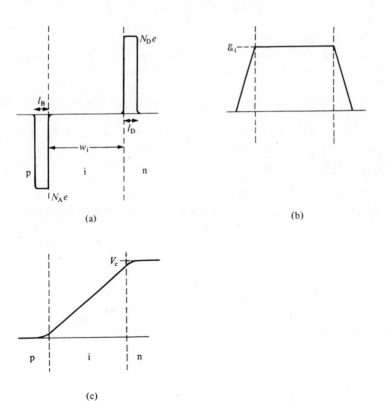

Figure 2.24. Depletion layer in a p-i-n diode: (a) charge profile; (b) field profile; (c) voltage profile.

The implication above is reasonable *ab initio*. On bringing into contact, for example, the p and i regions of semiconductor, the difference in hole densities ($p_p - p_i$) is much smaller than was the case in the p-n junction and fewer holes will diffuse across the junction.

The first practical point about the p-i-n diode is that given voltages are associated with much lower internal field strengths. Although the same field strength is necessary to initiate breakdown in both cases, nevertheless, in the p-i-n junction, a very much larger applied voltage is necessary to produce it. Thus, p-i-n diodes are used in high-voltage applications.

Since less charge is associated with a given applied voltage, there will be smaller changes in charge for given changes in applied voltage. As $\Delta Q$ equals $C\Delta V$, it follows that the capacitance is much smaller than in the p-n diode. In practice, the p-i-n depletion layer capacitance is of the order of 1 pF.

This might suggest an improved high-frequency response, but, in fact, the p-i-n diode has a low cut-off frequency. The great width of the intrinsic depletion layer means a larger excess charge when it is forward-conducting. On reverse biassing, the excess charge can support a reverse current for a considerable time. In other words, the p-i-n diode has a particularly large diffusion capacitance (Kleinman, 1956). The overall capacitance characteristic of a p-i-n diode is shown in Figure 2.25.

The p-i-n diode may be used rather more effectively than the simple p-n diode as a microwave switch, because of its lower reverse capacitance, and its higher breakdown voltage. However, an additional use of the p-i-n diode arises from the fast edge generated in the current waveform consequent on a voltage reversal in the external circuit.

When the diode is conducting, considerable charge is stored in the diffusion capacitance, although the voltage $V_j$ is quite small. The equivalent circuit is shown in Figure 2.26a, where the effective d.c.

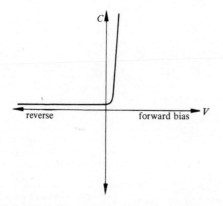

**Figure 2.25.** Capacitance characteristic of a p-i-n diode.

resistance $R_d$ of the diode is usually negligible compared with that of the external circuit, $R_{cct}$. If $V_B$ is now reversed, the equivalent circuit, neglecting leakage current, is as shown in Figure 2.26b. The junction voltage will not change until the diffusion capacitance is discharged so that a reverse current flows, almost equal to the original forward current.

The reverse current steadily discharges the diffusion capacitance until the capacitance itself becomes negligible. During the discharge, the diode voltage $V_J$ has fallen only slightly, from about 0·7 V to about zero. At this stage, since $I$ equals $C dV/dt$ and $C$ is now extremely small (the depletion capacitance is of the order of 1 pF), the junction voltage rapidly assumes its final equilibrium value, the current dropping simultaneously to zero (Moll *et al.*, 1962).

Because of the low value of the depletion layer capacitance, this transition occurs extremely rapidly, typically between 100 ps and 1 ns. The resultant current wavelength is therefore very rich in harmonic content, and the p–i–n diode may be used as a microwave generator at several gigahertz when driven with a fundamental frequency of a few hundred megahertz.

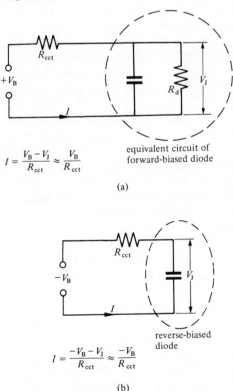

$$I = \frac{V_B - V_J}{R_{cct}} \approx \frac{V_B}{R_{cct}}$$

equivalent circuit of forward-biased diode

(a)

$$I = \frac{-V_B - V_J}{R_{cct}} \approx \frac{-V_B}{R_{cct}}$$

reverse-biased diode

(b)

**Figure 2.26.** p–i–n diode circuit: (a) forward-biassed; (b) reverse-biassed.

In principle, tunnel diodes could be used in a similar way but, in practice, their low voltage and current ratings over the negative-resistance region cannot compete with the p-i-n diode which has a forward current of hundreds of milliamperes and a reverse breakdown voltage of tens of volts.

(The rather low reverse breakdown voltage for p-i-n diodes in the application quoted above may appear surprising in view of the earlier statement that they can be used in high-voltage applications. In the latter case a very wide intrinsic layer would be employed, but, for harmonic multiplication, where the fundamental may be some hundreds of megahertz, a narrower intrinsic layer is used so that the diffusion capacitance is sufficiently small for discharge during the 'off' half-cycle.)

### 2.8 IMPATT diodes

The desirability of generating microwaves, even at low power by solid-state devices, cannot be overestimated. Until such devices were developed, the only sources of microwaves were the klystron for low power (around a watt) and the magnetron for up to megawatts of pulsed power. These are physically large, heavy devices, and even the low-power klystron requires power supplies of some hundreds of volts; they are certainly not compatible with the miniaturisation of modern semiconductor circuitry.

The basis of the IMPATT diode [1] is a p-n junction reverse-biased to just below the avalanche breakdown threshold, and arranged so that there is a specific time lag before the external current responds to any subsequent avalanching.

Consider an ordinary p-n junction, reverse-biased as above, and super-impose a single cycle of additional reverse bias (see Figure 2.27a). Avalanche breakdown will occur, and the charge carriers within the depletion layer will increase in number throughout the positive half-cycle. During the negative half-cycle the excess charge carriers will decay, as in Figure 2.27b. The excess carriers will move across the depletion layer and there will be a further delay, due to their finite velocity, before the external current responds as in Figure 2.27c.

The effect of superimposing a continuous sine wave of appropriate frequency onto the d.c. bias will be a current waveform which, although distorted, will be fundamentally 180° out of phase with the input voltage. In conjunction with an external tuned circuit, the negative-resistance characteristic leads to continuous oscillation.

The excess charge generated in the abrupt p-n junction during the avalanche is limited, since the electric field will only exceed the threshold value over a comparatively narrow region of the depletion layer, Figure 2.2b. The resulting changes in current are rather small, and the negative resistance ($R = dV/dI$) tends to be rather large.

[1] IMPact, Avalanche, Transit Time diode.

A method of increasing the avalanche breakdown is to include an intrinsic layer between p and n regions (the p-i-n diode again). The uniform field in the intrinsic layer ensures that avalanching occurs over more or less the full width of the depletion layer.

The Read diode is a more complex device. It has a four-layer p-n-i-n configuration in which the avalanche carrier generation and the subsequent transit time delay are physically separated. The field distribution is shown in Figure 2.28, although a flat top may be built into the peak if the p material is prepared by diffusion into the n (i.e. a graded junction). Avalanche generation occurs only within the p-n junction region, whereas almost the whole of the delay occurs within the intrinsic region.

By separating the generation and delay, the output current pulse is very much neater than in the simpler devices, and the overall efficiency of the Read diode is better. On the other hand, it is naturally more difficult to manufacture.

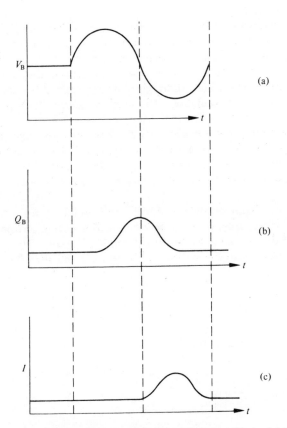

**Figure 2.27.** Action of the IMPATT diode. Plots of (a) applied voltage, (b) depletion layer charge, and (c) external current against time.

In practice, IMPATT diodes may be used to generate microwave power of about a watt or more, at frequencies up to the tens of gigahertz range (Lee *et al.*, 1968).

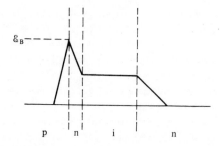

**Figure 2.28.** Field profile in the Read diode.

### 2.9 The Schottky diode

The last device to be discussed in this chapter is a metal semiconductor junction, a metallic conductor being separated from a semiconductor by an insulating film (Schottky, 1942). As carriers must tunnel through this film, it is kept as thin as possible, and, in practice, the inevitable oxide film on the semiconductor surface, although normally negligibly thin, is often directly suitable as the insulator. The metallic coating is then simply deposited on top.

Suppose that the semiconductor has a lower work function than the metal; then there will be a net flow of electrons from the semiconductor to the metal. Equilibrium is reached when the negative potential of the metallic conductor is sufficient to prevent further accumulation of electrons.

The excess negative charge on the metal will reside on its surface, i.e. the metal-insulator interface, whereas the effective positive charge on the semiconductor will diffuse some way into the semiconductor. The resultant excess charge density is shown in Figure 2.29 and the effect on

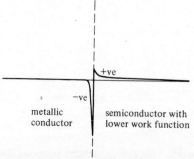

**Figure 2.29.** Charge distribution in a Schottky diode.

the energy bands is shown in Figure 2.30 (Shockley and Pearson, 1948) (compare the MOSFET).

As in the p-n junction, electrons are continually crossing and re-crossing the junction, and in the absence of external current flow $I_m$ must equal $I_{sc}$ (see Figure 2.31a).

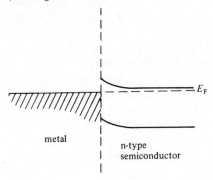

**Figure 2.30.** Energy diagram of Schottky diode.

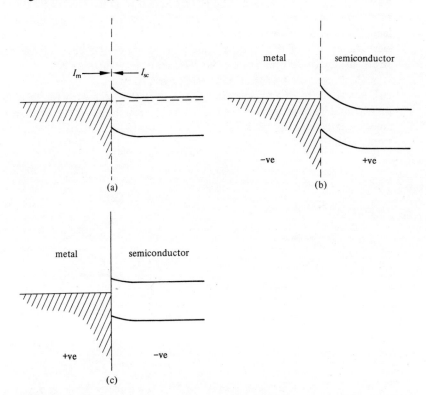

**Figure 2.31.** The effects of bias on the Schottky diode: (a) zero bias; (b) metal negative; (c) metal positive.

The characteristics of the metal–semiconductor junction become apparent on considering the effect of applied voltages on the two current components, $I_m$ and $I_{sc}$. If the semiconductor is made positive with respect to the metal, its energy levels will be lowered (Figure 2.31b), and $I_{sc}$ will decrease. Biassing the semiconductor negatively raises the energy levels (Figure 2.31c), and $I_{sc}$ increases rapidly. With both positive and negative bias, $I_m$ is unaffected.

Qualitatively, the behaviour bears resemblance to that of the p–n junction, and the d.c. characteristic reflects the similarity.

In practice, additional conduction may occur by tunnelling. If the semiconductor is heavily doped, the depletion surface layer will be very thin, as in Figure 2.32a, and electrons can tunnel through the barrier. Again, if the difference between the two work functions is of the order of the semiconductor band gap energy, the surface layer of the semiconductor will be inverted as in Figure 2.32b. It is then possible for 'hole' tunnelling to occur.

The particular merit of the Schottky diode is its very short turn-off time (Krakauer and Soshea, 1963). Conduction is by electron flow, where the electrons are majority carriers on both sides of the junction. In fact, the energy of the electrons arriving in the conductor must have at least the energy of the metal work function above the Fermi level; the electrons are therefore 'hot'. The relaxation time of majority carriers is very short, and that of 'hot' carriers is even shorter.

The advantage of the Schottky diode over the backward diode is that its current and voltage ratings are comparable with those of the ordinary p–n junction, whereas the backward diode is essentially limited to small-signal applications.

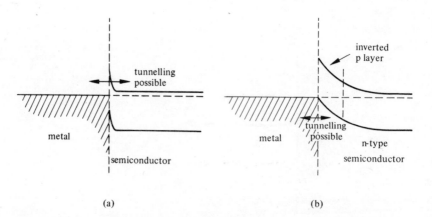

(a)                                                        (b)

**Figure 2.32.** Tunnelling possibilities on the Schottky diode: (a) heavily doped semiconductor; (b) large difference of work function.

## 2.10 Device materials

The most important physical parameter of any semiconducting material considered for general purpose junction devices is the width of its energy gap, $E_g$. The value of $E_g$ is important because it controls the intrinsic carrier level $n_i$ which in turn, by setting the likely ranges of carrier densities, determines bulk material resistivity, and junction I/V and temperature characteristics. Through $n_i$, $E_g$ also sets an upper limit to the natural impurity level acceptable in a semiconductor. This must be significantly less than likely doping densities (related to $n_i$) otherwise material characteristics may be influenced by the 'wrong' impurities.

Early devices were based almost entirely on germanium, $E_g = 0 \cdot 7$ eV, $n_i = 2 \cdot 5 \times 10^{19}$ m$^{-3}$, because it was readily available, could be made up into practical device format, and required not too unreasonable an upper limit to intrinsic impurity concentration ($\sim$1 part in $10^8$ or so).

Unfortunately the comparatively high value of $n_i$ in germanium leads to rather unsatisfactory temperature/leakage current characteristics. Minority carrier densities increase with temperature whereas majority carrier densities remain more-or-less constant, equal to the doping level, so that around a certain temperature the material loses its original p- or n-type character and junction devices cease to function as intended. Even before this stage is reached, the increased minority density leads to excessive leakage currents (reverse saturation) across pn junctions which seriously impair device performance.

As the critical temperature is primarily determined by $E_g$, silicon devices ($E_g = 1 \cdot 1$ eV) have the advantage over germanium that leakage currents at room temperature are normally quite negligible, or alternatively, that they may be usefully employed at temperatures up to $\sim$150°C, against $\sim$50°C maximum for germanium. The somewhat higher junction voltages for silicon ($\sim$0·7 V compared to $\sim$0·3 V for germanium) consequent on the lower carrier densities are sometimes a nuisance but are rarely a serious problem in practical circuitry. The initial difficulties associated with silicon technology have now been overcome, and silicon devices form the mainstay of present day semiconductor electronics.

For special purpose devices (see subsequent chapters) other semiconducting parameters may be more important and different materials preferable. For some applications high carrier mobilities are of prime importance whereas in others long carrier lifetimes are necessary. Unfortunately it seems to be an inevitable fact of life that these other characteristics are often found only in materials with some other shortcomings, for example, very low $E_g$, such that satisfactory device operation may be possible only in well refrigerated surroundings.

## References

Blakemore, J. S., de Barr, A. E., and Gunn, J. B., 1953, *Rept. Progr. Phys.*, **16**, 160.

Burrus, C. A., 1963, *Inst. Elec. Electron. Engrs., Trans. Microwave Theory Tech.*, **MTT-11**, 357.

Chang, J. J., Forster, J. H., and Ryder, R. M., 1963, *Inst. Elec. Electron. Engrs., Trans. Electron. Devices*, **ED-10**, 281.

Esaki, L., 1958, *Phys. Rev.*, **109**, 603.

Henderson, J. C., and Tillman, J. R., 1957, *Proc. Inst. Elec. Engrs. (London), Pt.B*, **104**, 318.

Kleinman, D., 1956, *Bell System Tech. J.*, **35**, 685.

Krakauer, S. M., and Soshea, S. W., 1963, *Electronics*, **36**, number 29, 53.

Lee, T. P., Standley, R. D., and Misawa, T., 1968, *Inst. Elec. Electron. Engrs., Trans. Electron Devices*, **ED-15**, 741.

McKay, K. G., and McAfee, K. B., 1953, *Phys. Rev.*, **91**, 1079.

Manley, J. M., and Rowe, H. E., 1956, *Proc. Inst. Radio Engrs.*, **44**, 904.

Manley, J. M., and Rowe, H. E., 1958, *Proc. Inst. Radio Engrs.*, **46**, 850.

Moll, J. L., Krakauer, S., and Shen, R., 1962, *Proc. Inst. Radio Engrs.*, **50**, 43.

Schneider, M. V., 1963, *Bell System Tech. J.*, **42**, 2972.

Schottky, W., 1942, *Physik*, **118**, 539.

Shockley, W., and Pearson, G. L., 1948, *Phys. Rev.*, **74**, 232.

Uhlir, A., 1958a, *Proc. Inst. Radio Engrs.*, **46**, 1101.

Uhlir, A., 1958b, *Proc. Inst. Radio Engrs.*, **46**, 1106.

Wolff, P. A., 1954, *Phys. Rev.*, **95**, 1415.

Zener, C., 1934, *Proc. Roy. Soc. (London), Ser.A*, **145**, 523.

# 3

# Transistors

During the twenty or so years since the point-contact transistor was first announced (Bardeen and Brattain, 1948), several other types of transistor have been successfully introduced.  These have been almost invariably junction devices, which offer much higher reproducibility in manufacture, and which are greatly superior in electrical performance.  In fact, the original point-contact transistor is now almost obsolete.

Devices described in this chapter include the basic junction transistor, the drift transistor, the unijunction, two types of field effect transistor, and silicon controlled rectifiers.  As the point-contact transistor is now a rarity and there was always doubt as to the exact *modus operandi,* there will be no further reference to it here.

A section is also devoted to what is surely a more significant development than that of any individual device, namely, the introduction of microminiaturisation and the production of integrated circuits.  Not only does this lead to extremely high component packing densities, it also alters the whole significance of circuitry design for the practising electronic engineer.

## 3.1 Junction transistors

Junction transistors are three-layer devices, basically either p-n-p or n-p-n sandwiches as in Figure 3.1 (Moll, 1955).  Both types function in the same manner, although potentials and charges, etc., will have opposite polarities in the two cases.  Thus, it is quite in order here to concentrate entirely on one or the other, and the n-p-n configuration has been chosen.

The main features of transistor operation may be appreciated by considering the three operating modes normally encountered in practice.

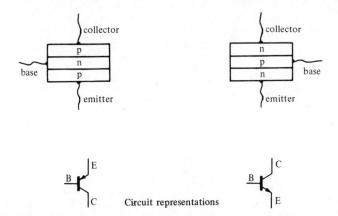

**Figure 3.1.** Schematic diagrams of n-p-n and p-n-p transistors.

In Figure 3.2, the voltage generator is such that the transistor base may be biased either positively or negatively with respect to the emitter.

First, suppose $V_{BE}$ to be negative. Both the transistor junctions, collector-base and base–emitter, will be reverse-biased and only the usual diode leakage currents will flow through them. These are commonly denoted as $I_{CBO}$ and $I_{EBO}$ respectively, and the total base current will be their sum. The carrier energy diagram of the transistor in this state is shown in Figure 3.3. In many applications, $I_{CBO}$ and $I_{EBO}$ will be negligibly small and to all intents and purposes the transistor behaves as a three-terminal open circuit. Not surprisingly, the transistor is said to be 'off'.

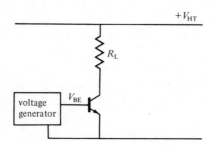

**Figure 3.2.** Transistor circuit for use in discussion of transistor characteristics.

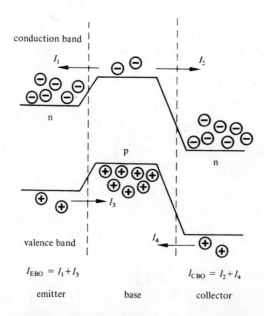

**Figure 3.3.** Energy band diagram of 'off' transistor.

Biassing the base positive with respect to the emitter will turn the transistor 'on'. If the positive bias is not excessive, the emitter junction will be forward-biased, whereas the collector junction will still be reverse-biased.

In practice, the emitter doping is made substantially heavier than the base doping, so that forward currents across the emitter-base junction are almost entirely due to carrier injection from emitter to base rather than from base to emitter (Chapter 2).

The carrier energy diagram of the transistor in this state (Shockley *et al.*, 1951) is shown in Figure 3.4. As shown in Chapter 2, $V_{BE}$ will determine the minority carrier density (electrons) $n_{pX}$ on the base side X of the emitter-base depletion layer.

$$n_{pX} = n_p \exp\left(\frac{eV_{BE}}{kT}\right).$$

Similarly, by considering the base-collector depletion layer, the electron density at Y is

$$n_{pY} = n_p \exp\left(\frac{eV_{BC}}{kT}\right).$$

Again these densities will be maintained despite any modest current flow that might ensue.

The base region, being rich in majority carriers, is essentially field-free, but, because of the higher minority carrier density at X than at Y, a diffusion current will cross the base from the emitter towards the collector.

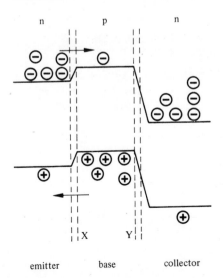

Figure 3.4. Energy band diagram of 'active' transistor.

If the device geometry is favourable (Figure 3.5), and loss of carriers through recombination is minimised by having very narrow base widths, then almost all the emitter-to-base current will arrive at the collector junction, and will cross into the collector. The fraction of emitter current crossing into the collector is denoted by $\alpha$, which in practice usually has a value between $0 \cdot 95$ and $0 \cdot 997$.

As the minority carrier flow across the base is more or less constant throughout, it follows from the diffusion equation, $I = eD\,\mathrm{d}n/\mathrm{d}x$, that the minority carrier density within the base must decrease linearly from X to Y, as shown in Figure 3.6.

With a narrow base width, recombination does not occur to any great extent, but it is nevertheless of great significance since it leads to an external base current flowing sufficient to maintain the equilibrium majority carrier density.

The base current will be equal to the net recombination rate of excess minority carriers within the base, which itself will be proportional to the total excess base charge, $Q_B$.

Now, since $I = eD\,\mathrm{d}n/\mathrm{d}x$, it follows that $Q_B = \displaystyle\int I\,\mathrm{d}x/D$, integrated over

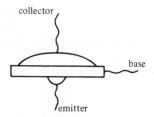

**Figure 3.5.** Cross section of practical alloy junction transistor.

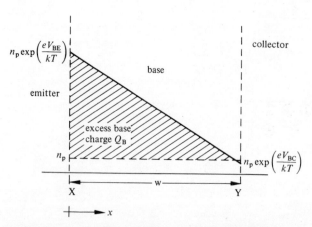

**Figure 3.6.** Base charge distribution in active transistor.

the base width. Thus, $Q_B = Iw/D$, where $w$ is the base width, and $I$ is taken as the collector current, so that finally one may write that $I_C$ is directly proportional to $I_B$. The ratio $I_C/I_B$ is usually denoted by $\beta$: $I_C = \beta I_B$. Since $I_B = I_E - I_C$, it follows that

$$I_C = \frac{\beta}{\beta+1}I_E , \quad \text{or} \quad \alpha = \frac{\beta}{\beta+1} ,$$

showing that $\alpha$ is also a constant for a given device irrespective of the current amplitudes. As $\alpha$ is close to unity, $I_C$ will be greater than $I_B$ and this provides the basic amplifying feature of the transistor. Typical values of $\beta$ lie between 20 and 300.

Operated in this mode, a transistor is said to be in the active, or linear, region of its characteristics, and this is, of course, the mode chosen for orthodox sinusoidal amplification. The input is usually applied to the base, and the output taken from the collector. The input may be considered either as a voltage signal controlling $V_{BE}$, or as a current signal controlling $I_B$. In practice the linearity of the $(I_B, I_C)$ characteristic as opposed to the exponential $(V_{BE}, I_C)$ characteristic (Figure 3.7) leads to the design of circuits often being tackled from a consideration of currents rather than voltages.

If the base drive is increased, the collector current will also increase, and the collector potential will fall ($V_{CE} = V_{HT} - I_C R_L$). With sufficient base drive, the fall in collector potential will be such that the collector junction loses its reverse bias ($V_{CE} = V_{BE}$). The value of $I_C$ at which $V_{CB}$ is zero is $I_C = (V_{HT} - V_{BE})/R_L$, or as $V_{BE}$ is usually small in comparison with $V_{HT}$, $I_C = V_{HT}/R_L$. The corresponding value of base current is $I_B = V_{HT}/\beta R_L$. Further increase in $I_B$ beyond $V_{HT}/\beta R_L$ results in the third mode of operation, where both emitter-base junction and collector-base junction are forward biased.

In this state, carriers will be injected from the collector to the base and the net collector current will be the difference between the diffusion current, still crossing the base from emitter to collector, and the additional

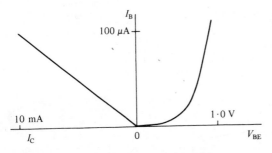

**Figure 3.7.** Typical transistor characteristic.

collector-to-base current, diffusing across the base from collector to emitter (see Figure 3.8).

Despite the injection of additional carriers from collector to base, the net collector current direction cannot be reversed since the collector lead is returned to a positive potential. This means that the net minority carrier density across the base must still decrease from the emitter towards the base, and, therefore, $n_{pX} > n_{pY}$, whence $V_{BE} > V_{CE}$. In other words, the collector potential will remain positive, somewhere between $V_{BE}$ and zero.

For most practical purposes, $V_{CE}$ (less than $V_{BE}$) may be neglected in comparison with $V_{HT}$, so that the collector current 'saturates' at $I_C$ equal to $V_{HT}/R_L$, however much $I_B$ is increased above $V_{HT}/\beta R_L$. The transistor is said to be saturated in this mode of operation. (An alternative expression is that the transistor is 'bottomed', originating from the fact that its operating point has reached the bottom end of the linear region of the characteristics.) The saturation of the collector current is reflected in the constant gradients of base charge shown in Figure 3.9.

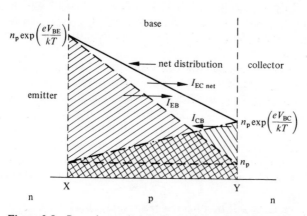

**Figure 3.8.** Base charge distribution in saturated transistor.

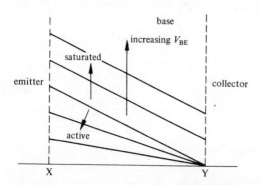

**Figure 3.9.** Variation of base charge distribution with increasing $V_{BE}$.

It is interesting to note in passing that, despite the collector–base junction being forward-biased, the current that flows is, in fact, a substantial reverse current. This confirms the earlier statement (Chapter 2) that junction polarities do not *directly* determine current directions, let alone their magnitudes.

In saturation then, $V_{BE}$ increases beyond the value at which $V_{CB}$ falls to zero, and carrier injection, from collector to base, results. A heavier build-up of base charge occurs, and minority carrier recombination increases. This increase is enhanced by the geometry of the transistor (Figure 3.5) which reduces the proportion of collector-to-base carriers likely to arrive at the emitter. A much higher proportion of this current, therefore, recombines within the base, and $I_B$ rises rapidly. As $I_C$ has saturated, $I_B$ will rise at the same rate as $I_E$, i.e. at a rate appropriate to a simple forward-biased junction (Figure 3.10).

In the saturated state, the emitter, base, and collector potentials are all quite close to each other. As a reasonable first approximation, the equivalent circuit of a saturated transistor may often be taken as a direct connection between the three terminals, just as if the transistor were a blob of solder. Bearing in mind that the equivalent circuit of the 'off' transistor simply has open circuits between the emitter, base, and collector, it will be appreciated that the 'off' and saturated states form the basis of the use of the transistor as a switch (Sparkes, 1960). With $V_{BE}$ zero or negative, $I_C$ is zero; with $V_{BE}$ sufficiently positive to saturate the transistor, $I_C = V_{HT}/R_L$. The transistor is very effective as a switch, since in the 'off' state the leakage current through the load may be well below a microamp, whereas in the 'on' state, all but a few tenths of a volt of the supply voltage will appear across the load. A particularly valuable feature of the transistor switch is that it may be switched repeatedly at rates higher than 10 MHz if necessary.

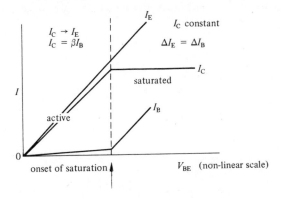

**Figure 3.10.** The variation of transistor currents with $V_{BE}$.

## 3.2 Junction transistors at high frequencies

There are various factors that contribute to the fall in $\beta$ at higher frequencies that is observed for bipolar transistors. Most of these may be ascribed to various capacitive aspects of transistor action.

For instance, both junctions have depletion capacitance. At high frequencies the emitter–base depletion capacitance will shunt the input signal, whereas the collector–base capacitance may both shunt the output signal and allow feedback direct from collector to base. Of these capacitances, that associated with the collector is the more serious, because of the feedback possibility and because it usually appears in higher-impedance surroundings.

However, as in the single-junction diode, diffusion capacitance is also of great significance. The excess base charge gives 'inertia' to the collector current, which has to be overcome before changes in input can affect the output current.

It was shown in Chapter 2 that depletion capacitance can be reduced by lowering the semiconductor doping levels. Unfortunately there is a serious complication if this method is used to lower the collector capacitance in a transistor. With smaller doping levels, changes in the depletion layer width for changes in $V_{CB}$ become greater, and, as will be seen from Figure 3.11, this means that variations in the collector potential will have a significant effect on $I_C$, even if $V_{BE}$ is constant. In other words, the output impedance of the transistor is reduced, and the power gain falls [this is known as the Early effect (1952)]. Carried to an extreme, the collector depletion layer could reach right across the base to the emitter junction. Excessively large currents would flow, and the transistor would be permanently damaged (punch-through).

A transistor made as suggested above might have a somewhat improved frequency response, but would suffer from a lower output impedance, and lower collector operating potentials to avoid punch-through. However, perhaps rather surprisingly, it was found that attempts to reduce the base

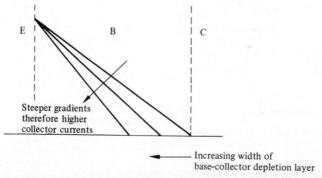

Figure 3.11. The effect of collector potential variation on effective base width.

diffusion capacitance also brought about an improvement in the collector depletion capacitance.

The minority carrier density, at such a position as at X in Figure 3.6, can follow almost instantaneously any variation in $V_{BE}$, but it is not until the density near position Y has responded that either $V_{CE}$ or $I_C$ will change. In other words, the entire minority charge distribution within the base must change (as in the active region of Figure 3.9) before the collector current and voltage can take up new equilibrium values.

A consideration of this point suggests various ways of speeding up the response of the transistor. An obvious one is by reducing the width of the base, since this will reduce the time for changes in the density at X to diffuse across to Y. Alternatively it might be possible in some way to reduce further the base charge associated with a given collector current in transistors already having a minimum base width. Also, it might prove possible to find new semiconducting materials having higher carrier mobilities and, therefore, higher diffusion constants with lower relaxation times.

The first attempts at producing high-frequency transistors were made by reducing the base widths. New methods of construction allowed base widths to be narrowed from some $10^{-1}$ mm to around $10^{-3}$ mm, corresponding to an increase in cut-off frequency from tens of kilohertz to tens of megahertz.

Considerable attention is nowadays devoted to new semiconductors that offer the prospect of higher carrier mobilities (Jenny, 1958). In this respect germanium is superior to silicon; the electron mobilities are $0 \cdot 36$ m$^2$ V$^{-1}$ s$^{-1}$ and $0 \cdot 16$ m$^2$ V$^{-1}$ s$^{-1}$ respectively. Even better things were expected of compound semiconductors such as gallium arsenide $(0 \cdot 45$ m$^2$ V$^{-1}$ s$^{-1})$. Unfortunately, these hopes have not yet been fulfilled, owing mainly to technological difficulties in preparing the material to the necessary specifications.

At present, the most successful outcome of several different approaches is the graded base, or drift transistor. Such transistors are available for use up to a gigahertz and, in development, for even higher frequencies.

The drift transistor is again a three-layer sandwich device, n-p-n or p-n-p, the critical feature being that the base doping is not constant throughout the base region. With the perfection of diffusion techniques, it became possible to produce transistors in which the doping density within the base varied by some orders of magnitude from the emitter side (high doping) to the collector, while still maintaining a very narrow base width (Tanenbaum and Thomas, 1956).

The immediate effect of the doping variation, while the transistor is isolated, is the diffusion of majority carriers from the emitter side of the base towards the collector (see Figure 3.12). An electric field thus builds up, until, in equilibrium, there is no further net movement of carriers. It will be noticed that this field is in the appropriate direction to accelerate carriers, injected from the emitter, towards the collector. As is the intention, the transit times are substantially reduced.

In practice, the variation in doping across the base is roughly exponential, and the electric field strength is approximately constant, falling to zero at the collector junction. With a suitably steep doping gradient, the internal field will be sufficiently high for electron drift to be the dominant conduction mechanism, except near the collector junction where diffusion takes over. This means that the minority carrier gradient will be zero, except near the collector where it must be steep enough to support the collector current diffusion, exactly as in the basic junction transistor. The resultant base charge distribution will be as shown in Figure 3.13.

If a uniformly doped transistor were to carry the same collector current, the minority carrier distribution would have to be as shown by the dashed line in Figure 3.13. Clearly, the base charge associated with a given collector current has been significantly reduced, which is another desirable feature.

The graded doping, with a low density near the collector, results in a low collector depletion capacitance, but the risk of punch-through is reduced, since, as the depletion layer advances into the base towards the emitter, it encounters heavier and heavier doping, and its progress is correspondingly slowed down.

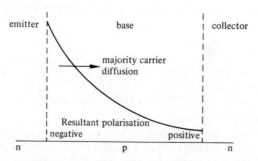

**Figure 3.12.** Net diffusion of majority carriers in a drift transistor.

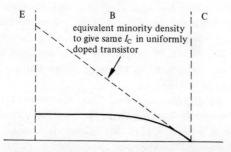

**Figure 3.13.** Minority carrier distribution in active drift transistor.

## 3.3 The unijunction

The basis of the unijunction transistor is the modulation of semi-conductor resistivity by the injection of additional carriers (Lesk and Mathis, 1953).

A schematic diagram of a unijunction is given in Figure 3.14. The p-type emitter is heavily doped compared with the n-type channel so that the effect of forward biasing the p-n junction is to inject holes from the emitter into the channel.

The operation may be described conveniently with the aid of Figure 3.15. With the emitter voltage zero, the emitter-channel junction is reverse-biased and the emitter current very small. The channel potential in the region of the emitter is approximately a fraction $l_1/L$ of the base 2 potential.

If the emitter voltage is steadily raised, there will be no change in channel conditions (other than a change in depletion layer width, which is not relevant here) until the emitter-channel junction becomes forward-biased. At this stage, holes will be injected into the channel from the emitter, and these holes will move towards base 1.

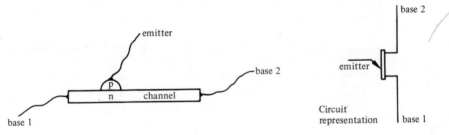

**Figure 3.14.** Schematic diagram of unijunction transistor.

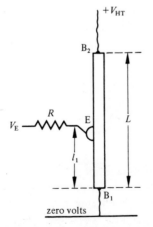

**Figure 3.15.** Circuit for use in discussing unijunction operation.

There are now more carriers in the emitter-base 1 part of the channel, so its conductivity is higher than it was originally. Consequently, the channel potential around the emitter falls below the original value, and the forward bias of the emitter-channel junction increases.

Clearly the increase in conductivity is cumulative, and care must be taken to prevent excessive current flow from damaging the device. This is usually effected by limiting the current by internal resistance in the emitter source. As the emitter current increases, its potential falls, and the emitter-base 1 characteristic displays a region of negative resistance.

The well-defined threshold voltage at which the cumulative action begins, as well as the negative-resistance emitter characteristic, makes the device particularly suitable for use in relaxation oscillators. A typical circuit is shown in Figure 3.16. Initially the capacitor is uncharged and $V_E = 0$. The capacitor charges exponentially with time through $R$ until $V_E$ reaches the unijunction turn-on potential. The resistance between the emitter and base 1 then falls very rapidly, allowing $C$ to discharge through the unijunction, and consequently $V_E$ falls. The limit to the decay is reached when $V_E$ is so low that any further fall must reduce the forward bias on the emitter-channel junction. At this stage, the channel conductivity begins to fall and cumulative action now goes rapidly in the opposite direction, finishing with $V_E$ still very low (held down by the discharged capacitor), and the channel potential in the emitter region back at its initial high value. The emitter-channel diode is reverse-biased, and, therefore, the emitter current is zero. The capacitor again begins to charge through $R$, and the whole cycle repeats itself.

The unijunction is not suitable for normal amplification purposes.

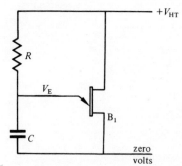

**Figure 3.16.** Unijunction relaxation oscillator.

## 3.4 Field effect transistors

Junction transistors are undoubtedly the backbone of present electronics, but nevertheless they do have characteristics which are often disadvantageous. The low input impedance (essentially that of a forward-biased junction) is troublesome in high-impedance work, and the fact that they are minority carrier devices renders them sensitive to any disturbance that is likely to alter the minority carrier concentrations. It is well known that changes in temperature or exposure to atomic radiation upset their operation. The more recent field effect transistor (FET) is superior in these respects, and having other valuable characteristics of its own in addition, is already widely in use.

The basic principle of the FET is, in fact, similar to that of the unijunction, namely the modulation of the overall conductivity of a sample of semiconductor. Whereas in the unijunction this is achieved by varying the number of carriers present, in the FET it is the actual width of the conducting channel that is varied (Eimbinden, 1964).

There are two types of FET, differing in the manner in which the channel width is controlled. In the junction-gate FET, variation of a gate electrode potential changes the width of the depletion layer of a reverse-biased junction, the depletion layer extending into the conducting channel. In the metal-oxide-semiconductor FET (MOSFET) variations in the potential of an electrode near the semiconductor surface, but insulated from it, alter the character of that semiconductor to a varying depth below the surface, and again result in the channel width being affected.

In both types of FET it is the behaviour of majority carriers which determines the electrical characteristics, and the advantages of avoiding minority carrier control should follow directly.

### 3.4.1 The junction FET

A schematic diagram of the junction FET is shown in Figure 3.17. In the absence of any applied potentials, there will be the normal high-resistivity depletion layer which will already partially restrict the conduction channel width between source and drain.

If the gate is now negatively biased with respect to the channel as in Figure 3.18, the depletion layer will move further into the channel. With sufficient reverse bias $V_P$, the depletion layer will completely block the channel, which is then said to be 'pinched off'.

The pinch-off voltage $V_P$ may be found by applying a previous result from Chapter 2, that relates to the depletion layer width, that is

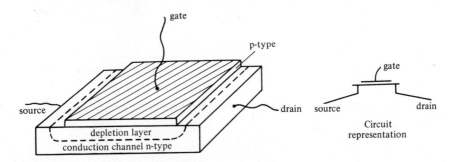

**Figure 3.17.** Schematic diagram of junction-gate FET.

$V_P = \pm eNW^2/2\epsilon$, where $W$ in this case is the overall width of the n-type material, and $V_P$ is positive for p-channel devices, negative for n-channel.

Clearly, then, variations in gate potential will affect the source-drain conductivity. However, the width of the depletion layer is also affected by the drain current $I_{DS}$. This follows since it is the potential difference between gate and channel that determines the depletion layer width, and, with a drain current flowing through the channel, the potential difference will vary from one end of the gate region to the other.

Figure 3.19 shows the effect on the depletion layer of varying $I_{DS}$ at constant $V_{GS}$. Thus, even if the device is not initially pinched off, it may be made to pinch off by increasing $V_{DS}$ and hence increasing $I_{DS}$. By considering the drain end of the channel at pinch-off, it is apparent that the pinch-off drain potential $V_{DSP}$ is given by $V_{GS} - V_{DSP} = V_P$ (in our example, $V_P$ and $V_{GS}$ are negative quantities, and $V_{DSP}$ is positive).

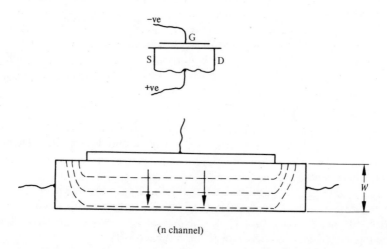

**Figure 3.18.** Change in depth of depletion layer with increasing $V_{GS}$ ($V_{DS} = 0$).

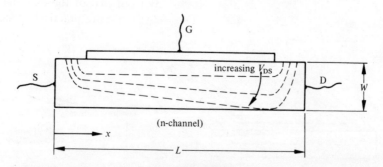

**Figure 3.19.** Change in depletion layer profile with increasing $V_{DS}$.

A common analysis of the active region (not pinched off) assumes the geometry shown in Figure 3.19 (Shockley, 1952). The channel potential at any point $x$ referred to the source is denoted $V_{CS}$. The potential across the gate-channel junction is therefore given by

$$V_{GC} = V_{GS} - V_{CS} .$$

Now the depletion layer width at $x$ is equal to $[2\epsilon(-V_{GC})/eN]^{1/2}$ so that the conduction channel width at $x$ is $\{W - [2\epsilon(-V_{GC})/eN]^{1/2}\}$. Thus, the potential drop along a channel element length $dx$ is given by

$$dV_{CS} = \frac{I_{DS}\rho\,dx}{W - [2\epsilon(-V_{GC})/eN]^{1/2}}$$

or

$$I_{DS}\rho\frac{dx}{dV_{CS}} = W - \left[\frac{2\epsilon}{eN}(V_{CS} - V_{GS})\right]^{1/2},$$

where $\rho$ is the channel resistivity. Substituting $V_P = -eNW^2/2\epsilon$,

$$I_{DS}\rho\frac{dx}{dV_{CS}} = W\left[1 - \left(\frac{V_{GS} - V_{CS}}{V_P}\right)^{1/2}\right].$$

Integrating along the length of the channel, from $x = 0$ to $x = L$, we find

$$I_{DS}\rho L = W\left[V_{CS} + \frac{2V_P}{3}\left(\frac{V_{GS} - V_{CS}}{V_P}\right)^{3/2}\right],$$

the limits to the right-hand side being $V_{CS} = 0$ and $V_{CS} = V_{DS}$. Thence,

$$I_{DS} = \frac{W}{3\rho L}\left[3V_{DS} - 2\left(\frac{V_{GS}}{V_P}\right)^{3/2}V_P + 2\left(\frac{V_{GS} - V_{DS}}{V_P}\right)^{3/2}V_P\right]$$

or

$$I_{DS} \propto \left[\frac{3V_{DS}}{V_P} - 2\left(\frac{V_{GS}}{V_P}\right)^{3/2} + 2\left(\frac{V_{GS} - V_{DS}}{V_P}\right)^{3/2}\right].$$

The above derivation neglects the junction contact potential, by which the actual junction potential differs slightly from that of the external gate connection. Also, the geometry is not typical, a more practical version

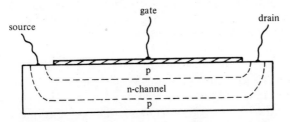

**Figure 3.20.** Schematic diagram of diffused junction-gate FET.

would be as shown in Figure 3.20. More serious is that devices are usually prepared by diffusion of impurities from the material surface (see Chapter 3.6), and therefore the doping density $N$ varies across the channel width. Not surprisingly, the analysis becomes more complex if these factors are taken into account, and its presentation here is not warranted.

What happens to the channel if $V_{DS}$ is increased above pinch-off? As pinch-off is approached, the conduction channel becomes very narrow and high field strengths will occur. Under these conditions the carrier velocities tend to saturate (Grosvalet *et al.*, 1963). This means that any further narrowing of the channel would reduce $I_{DS}$, but this cannot occur with increasing $V_{DS}$, as lower $I_{DS}$ would mean lower potentials along the channel from the source, and, hence, a widening of the channel. The effect of increasing $V_{DS}$ above pinch-off is that the channel width remains constant within the pinched-off region, and $I_{DS}$ settles to a more or less constant value (saturation).

As $I_{DS}$ is more or less constant in saturation, there will be no significant changes in the potential distributions within the channel, on either the source or the drain side of the pinched-off region. Any variations in $V_{DS}$ must then be taken up by variations in the length of this region.

The above equation, relating $I_{DS}$ to $V_{GS}$ etc., could be used to determine the saturated drain current $I_{DSP}$ by substituting the value of $V_{DS}$ at which pinch off just occurs, that is $V_{DS} = V_{DSP} = V_{GS} - V_P$. However a more accurate expression, relevant to practical devices, Figure 3.20, is that

$$I_{DSP} \propto \left(\frac{V_{GS}}{V_P} - 1\right)^2.$$

The square law relationship between $V_{GS}$ and $I_{DSP}$ is well observed in practice and is deliberately employed, for example, in analog multiplication by the quarter-squares method, or for square law detection.

In practice, the onset of carrier velocity saturation is gradual, and the increase in $I_{DS}$ with $V_{DS}$ flattens off comparatively gently in passing from the active to the saturated region. A typical set of n-channel junction FET characteristics is shown in Figure 3.21; characteristics of p-channel FET's are similar only with current and voltage polarities reversed.

As normally used, the gate connection notices only a reverse-biased junction, so the gate currents are typically of the order of nanoamperes. Coupled with gate potentials of some volts, this implies an input resistance of the order of $10^9$ $\Omega$. Clearly, the high gate resistance will be lost if the gate-channel junction becomes forward-biased anywhere along its length. and this means in practice that $V_{DS}$ and $V_{GS}$ must have specified, opposite, polarities.

Not surprisingly, the gate current is usually quite insignificant in practice, and the drain current may be considered to be controlled purely by the gate potential (shades of the thermionic valve!). This means, then, that a useful parameter of the FET for normal amplifying purposes is its mutual conductance $g_m$, where $g_m = dI_{DS}/dV_{GS}$. Since

$$I_{DS} \propto \left( \frac{V_{GS}}{V_P} - 1 \right)^2 ,$$

it follows that

$$g_m \propto \left( \frac{V_{GS}}{V_P} - 1 \right) \propto (V_{GS} - V_P) ,$$

or

$$g_m \propto (I_{DS})^{1/2} .$$

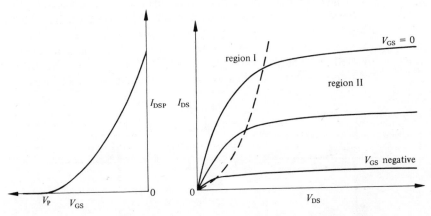

**Figure 3.21.** Electrical characteristics of n-channel junction-gate FET.

### 3.4.2 The MOSFET or MOST

An alternative method of modulating the width of a conducting channel depends on the effect of an electric field on the surface region of a semi-conductor. Previously we have been concerned entirely with the bulk properties of semiconductors, but we must now consider what happens at the surface.

At the surface of a tetravalent semiconductor there must be unfilled valence bonds, in fact, two per atom. These bonds act as electron traps such that mobile electrons moving into their vicinity will be held by them. The result is that on an ideal semiconductor surface there is a sheet of negative charge (Handler, 1964).

The effect of the negative charge is to repel free electrons in the semi-conductor away from the surface and to attract holes towards the surface. In other words, the surface has a tendency to be more p-type (or less n-type) in character than the bulk material. Looked at in another way, any free electrons near the negatively charged surface are in a higher energy environment than similar electrons in the bulk of the material, indicating an upward bending of the energy levels near the semiconductor surface.

Figure 3.22a shows the charge densities and energy bands associated with the surface of a lightly doped n-type semiconductor. If the doping is sufficiently light, the holes attracted towards the surface by the negative charge may outweigh the electrons in that region, resulting in a layer of the material at the surface showing p-type properties. Under these conditions, the surface layer is said to be inverted, the inverted layer being separated from the bulk material by a depletion region. If the bulk material is heavily doped (Figure 3.22b) the surface charge will be insufficient to produce inversion.

A charged metal electrode, on the surface of the semiconductor but insulated from it, may be used to modify the effect of the inherent surface charge. If the electrode is negatively charged, the effect of the surface charge will be augmented and even a heavily doped semiconductor may be inverted at its surface. On the other hand, a positively charged electrode will oppose the surface charge, and an inverted layer will shrink.

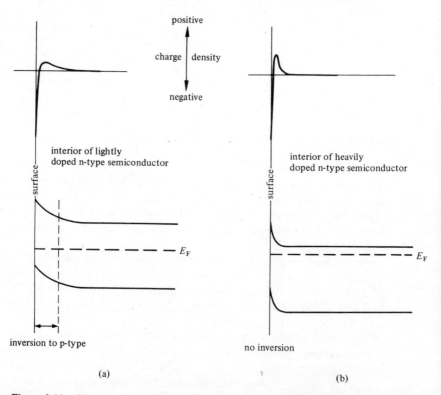

**Figure 3.22.** Change in properties of n-type semiconductor at surface: (a) lightly doped material; (b) heavily doped material.

Such behaviour is the basis of MOSFET action, two types of device being available, dependent on whether an inverted layer is initially present (lightly doped substrate) or not (heavily doped substrate). Idealised arrangements are shown in Figure 3.23a and 3.24a.

Conduction between the drain and source is only possible if there is a continuous n-type channel between them. The device shown in Figure 3.24a will therefore conduct even with zero gate bias. Altering $V_{GS}$

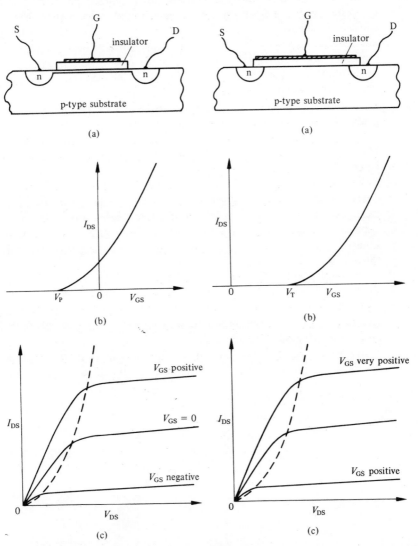

**Figure 3.23.** Depletion-type MOSFET: (a) basic device; (b) gate characteristics; (c) drain characteristics.

**Figure 3.24.** Enhancement-type MOSFET: (a) basic device; (b) gate characteristics; (c) drain characteristics.

will affect the width of the inversion layer and vary the position of the depletion layer. The device is said to be a 'depletion-mode' MOSFET. The device shown in Figure 3.24a will not conduct with zero gate bias. A minimum bias $V_T$ has to be applied to enhance the effect of the surface charge before a conduction channel is formed. The device is an 'enhancement' MOSFET.

Typical characteristics of the two devices are shown in Figure 3.23b, 3.23c, 3.24b, and 3.24c (characteristics of p-channel MOSFET's will have opposite current and voltage polarities). It can be seen that they are similar in principle to those of the junction-gate FET, showing both active and saturated (pinch-off) regions. However, the different biasing requirements of MOSFET's offer particular circuit advantages. The depletion device may be operated with no standing bias on the gate, whereas the enhancement device may be operated with the mean value of $V_{GS}$ equal in both magnitude and polarity to $V_{DS}$. The latter possibility leads to such simplified amplifier circuitry as shown in Figure 3.25.

The gate input resistance is some orders of magnitude greater than that of the junction FET, being determined by the deliberately insulating layer. Typical values are around $10^{12}$ $\Omega$.

For small $V_{DS}$ ($V_{DS} \ll V_{GS} - V_P$), the $I_{DS}$, $V_{DS}$ characteristics are approximately linear, implying a particular drain:source resistance. This resistance is set by $V_{GS}$, and MOSFET's find useful application as voltage controlled resistors. The voltage control at the gate is very effectively isolated from the controlled channel resistance by the high gate input resistance.

In practice, the insulating layer is formed by oxidising the semiconductor surface, and depositing the metallic gate electrode on top. A consequent difference in behaviour from that described above is that the surface charge is usually positive rather than negative, but clearly this will not affect the principle of operation.

A further practical point is that the inverted layer of the depletion device is usually introduced by suitable impurity diffusion, rather than by

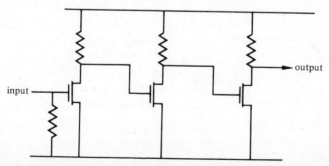

**Figure 3.25.** Simple three-stage MOSFET amplifier.

relying on the surface charge.  Surface charge densities are very susceptible to variations in surface preparation, purity of the insulating layer, etc., and a more reproducible device results from using the diffusion process.

The intrinsic upper frequency limit of the FET is very high (it is a majority carrier device) but, in practice, it is found that the frequency response is limited by 'external' factors.  Although the drain current responds extremely rapidly to changes in the applied voltage, the limiting factor is the rate at which these voltages can themselves be changed.  In practice, there is a considerable capacitance between the gate and the channel, acting as both capacitance across the input (gate-source) and as feedback capacitance (drain to gate).

The intrinsic frequency limit may be above a gigahertz, but, in practical circuits at present, the upper limits seem to be around hundreds of megahertz.

### 3.5 Thyristors

The ordinary p-n-p or n-p-n transistor is not normally suitable for use when currents of tens of amperes or more are involved.  Large collector currents mean correspondingly large base currents.  As these flow within a very thin base layer, sandwiched between the emitter and collector, a potential drop occurs in the base, so that the effective base-emitter potential falls off away from the actual base connection.  Figure 3.26 shows this effect in an n-p-n transistor, the potential at Y being less than that at X.  The result is that the emitter current concentrates itself in a comparatively small region near the base connection, and the remainder of the base-emitter junction hardly contributes (Emeis et al., 1958).

To handle larger currents there is little point in merely increasing overall junction area.  Improvements result from extended base connections (for example, a complete ring conductor circling the emitter) and, with the advent of more sophisticated preparation techniques, it is possible to build structures of the form shown in Figure 3.27.  Even so, the collector currents are still limited to amperes rather than tens of amperes or more.

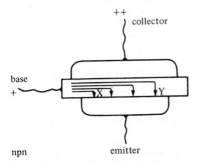

**Figure 3.26.** Variation of transverse base current in junction transistor.

The thyristor (Mackintosh, 1958) is a four-layer sandwich-type semi-conductor device, illustrated in Figure 3.28, wherein the whole junction areas are effective. Consequently thyristors are suitable for much larger currents than transistors, typically up to around 100 A.

In transistor operation, the predominant movement of charge carriers is from emitter to collector, but in the thyristor there are significant movements of both holes and electrons travelling in opposite directions. This makes the terms 'emitter' and 'collector' rather less apt, and thyristor electrodes are generally known as the anode, gate, and cathode (Figure 3.28).

The thyristor is not normally used in the active region, and it controls large currents by behaving as an on–off switch. Nevertheless, the gradual control of load currents is a straightforward application.

Use of the ordinary transistor as a switch has been mentioned earlier, but apart from its relatively low current-handling capacity, there is also the disadvantage in many situations that the base has to be driven all the time that one wants the load to be 'on'. In practice, it is often convenient to turn a switch 'on' by applying a momentary trigger pulse, with the switch then staying 'on' by itself. A thyristor behaves in this way.

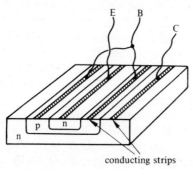

Figure 3.27. A possible arrangement for high-emitter current transistors.

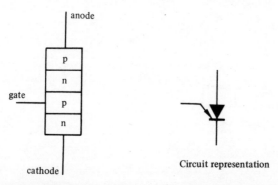

Figure 3.28. Schematic diagram of a thyristor.

Although the thyristor is a single device, some of its characteristics may be more readily appreciated by splitting it into a complementary p-n-p/n-p-n pair, as shown in Figure 3.29.

In Figure 3.29 there is a closed loop, which includes the gain of both transistors. It is therefore not surprising that the loop gain can exceed unity, resulting in infinite overall gain and both transistors saturating. In other words, $I_A$ can take finite values even with $I_G = 0$. This is the steady 'on' (saturated) feature shown by thyristors, the load current being limited by the load circuit external to the thyristor.

With $I_G = 0$, a second stable situation exists in which $I_A = 0$. The current gain $\beta$ of a transistor is not in fact constant, but falls off at both high and very low currents (Figure 3.30). As $I_A$ approaches zero, $\beta$ falls below unity and the overall gain of the system of Figure 3.29 falls from its infinite value to a more modest finite value. Thus, if $I_G$ is zero, $I_A$ can also be zero. This is the 'off' state of the thyristor. When $I_G$ is zero, the thyristor will remain in that one of its two stable states that is determined by its previous history.

The effect of triggering a thyristor from the 'off' state to the 'on' state is very similar to that which would be observed with the transistor pair. Provided that the trigger pulse is of sufficient amplitude for the $\beta$ product

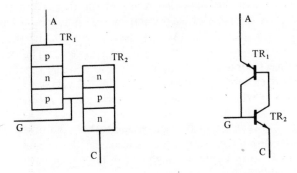

**Figure 3.29.** The representation of a thyristor by two directly connected transistors.

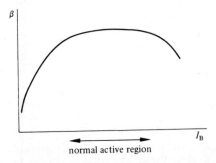

**Figure 3.30.** Variation of transistor $\beta$ with emitter current.

to be taken above unity, the thyristor will switch on, and it will then stay on after the trigger pulse is terminated and $I_G$ returns to zero. Clearly, for a given device, there will be a well-defined trigger threshold.

It is not usually practicable to turn a thyristor off at the gate. In effect it would be necessary to deflect the collector current of $TR_1$ away from $TR_2$ via the gate connection. As this current forms a considerable proportion ($\sim\frac{1}{2}$) of the thyristor load current, it will be appreciated that the gate circuitry would have to be capable of handling currents not much less than the actual load current.

The usual way to turn a thyristor 'off' is to reduce the anode potential momentarily to zero, or even to reverse it. Provided that sufficient time elapses so that the carrier densities fall to their equilibrium 'off' values, the thyristor will then stay 'off' even when the anode potential is raised to its positive value [1].

Although reducing the anode potential to zero, or even reversing it, may seem a somewhat impractical method turning a thyristor off, it is very well suited to the use of the device in controlling a.c. circuits where the polarity of the supply reverses once every cycle.

A simple control circuit is shown in Figure 3.31. The RC filter sets the trigger amplitude, and introduces a phase delay between the gate potential and the supply voltage. In this way, the 'firing' of the thyristor may be delayed as desired after the start of each input cycle. A typical set of waveforms is shown in Figure 3.32. As $V_{HT}$ rises from zero, the thyristor will stay off until $V_G$ becomes sufficiently positive to fire it. The device will then act as a direct connection between the load and the supply voltage until $V_{HT}$ falls to zero at $t_3$. By altering the RC values, the delay $t_2 - t_1$ may be varied at will, enabling a gradual variation of $I_L$ (mean) to be achieved.

As the thyristor is always either full on (anode–cathode voltage $\sim$ zero) or right off ($I_A = 0$), the power dissipation is low, and the overall efficiency of this method of control is high, much higher, of course, than it would be if a variable resistor were used.

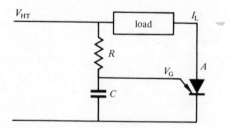

**Figure 3.31.** A simple thyristor control circuit.

[1] An important practical point is that there is a limit to the rate at which the anode potential is reapplied. If it is too fast, capacitive effects between anode and gate may result in an effective gate current, sufficient to retrigger the device 'on'.

A further multi-layer device is the triac (Howell, 1964), which might be described as a bi-directional thyristor. It is shown schematically in Figure 3.33, and can be seen to consist primarily of two four-layer devices in parallel, but with opposite polarities, and a single-gate control alongside.

The triac may be fired on every half-cycle of an alternating supply, whereas the thyristor can only fire on every other half-cycle. The thyristor is therefore useful where rectification is either desired or can be

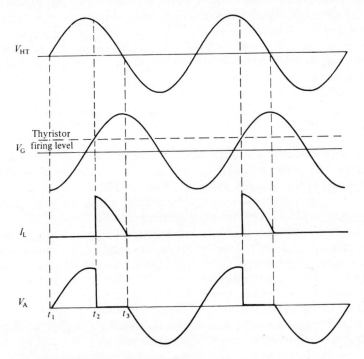

**Figure 3.32.** Thyristor control circuit waveforms.

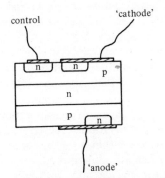

**Figure 3.33.** Schematic diagram of a triac.

tolerated; the triac is used where the load must have a genuine alternating current, or where the maximum r.m.s. power must be supplied.

## 3.6 Microminiaturisation

Individual junction transistors were first produced by taking small slices of n-type germanium, and by alloying trivalent metal contacts on both sides for the emitter and collector. A base connection was made to the slice itself. During manufacture, the device was heated to several hundred degrees Kelvin, at which temperature the emitter and collector 'blobs' melted and dissolved a little of the germanium. As the temperature was lowered, the germanium recrystallised with sufficient of the acceptor material impurity to outweigh the original donor concentration, and the p-n-p sandwich was formed.

The high-frequency response being determined essentially by the base width, the production of high-frequency transistors, implying very thin initial slices of semiconductor and the very precise control of the alloying process, was impracticable.

The breakthrough came with the controlled diffusion of impurities into semiconductor substrates, altering their character as desired, dependent on the valency of the impurity material. This led directly to the development of the drift transistor described earlier, and also paved the way for the subsequent production of microminiaturised circuits.

If a gaseous impurity is maintained at a steady concentration above a semiconductor surface, and the whole system is raised to a suitably high temperature, then impurity atoms will diffuse into the semiconducting material (Fuller and Ditzenberger, 1956). The depth of the diffusion is determined by the temperature, and by the duration of the diffusion process; careful control of these variables determines the impurity penetration very precisely. Figure 3.34 shows schematically the diffusion of acceptor impurities into an n-type semiconductor. The area of the diffusion can be controlled by suitable masking of the surface.

A further diffusion, but of donor impurities to a lesser depth, and covering a smaller area, results in the n-p-n structure shown in Figure 3.35. The base width may be made exceedingly narrow as well as most precise.

The complementary process is the epitaxial growth of a semiconductor by gaseous deposition. This is a necessary process, since it will be realised that, although successive diffusions can always alter a semiconductor from p- to n-type or vice versa, the concentration of impurity can only show an overall increase; the conductivity, thus, inevitably becomes greater and greater. Epitaxy offers the ability to deposit layers of lightly doped or intrinsic semiconducting material on top of a crystal irrespective of the crystal doping. The semiconductor is placed within a gaseous reaction system which precipitates silicon or germanium, as required. Condensation occurs on the semiconductor surface, and, with care, the

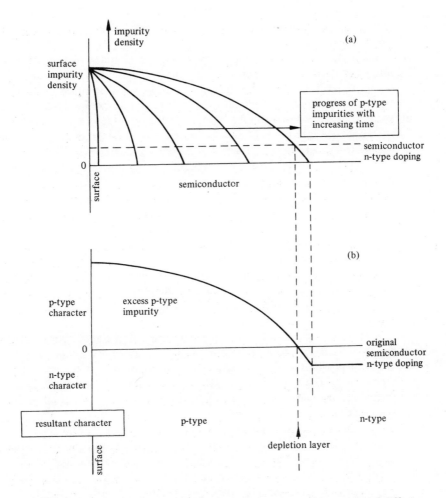

**Figure 3.34.** Diffusion of impurities into a semiconductor: (a) progress of diffusion with time; (b) resultant nature of the semiconductor at the end of the diffusion.

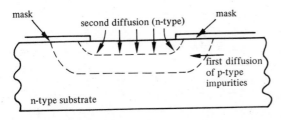

**Figure 3.35.** Formation of an n-p-n structure by successive diffusions, and masking.

continuity of the crystal lattice extends into the new material (Marinace, 1960; Lathrop, 1964).

Using these techniques, many transistors can be made on a single slice of semiconductor, in one operation, the individual devices being subsequently separated as desired. Other techniques allow the formation of insulating layers, by oxidising the silicon surface, and of conducting layers by the vacuum deposition of suitable metals. Clearly, MOSFET's, capacitors, resistors, and interconnecting conductors can all be produced on a semiconductor substrate. A range of components is shown in Figure 3.36.

A logical extension was the production of complete multi-transistor circuits on chips of semiconductor. In electronic engineering a great many complex systems can be made from a comparatively small number of basic circuits, e.g. multi-vibrators, amplifiers, logic gates, etc., and the prospect of integrating all the necessary components for these circuits into single pieces of semiconductor, often less than a millimetre square was obviously very exciting (Weimer et al., 1964). These hopes were in fact well founded. It is now possible for example to build a complete 'integrated' circuit on a silicon chip measuring considerably less than 2·5 mm square, whose equivalent in conventional components would need over 300 transistors and resistors.

A word or two should be added here about ion implantation. There are two disadvantages in preparing devices as above by impurity diffusion. It is not possible to prepare any arbitrary impurity profile, since the diffusion profile (e.g. Figure 3.34b) is rather limited, and further, when diffusing through a mask, the impurity diffusion spreads round underneath the edges of the mask as well as normally into the surface. This latter affects the precision with which small area diffusion can be achieved.

Ion implantation is an alternative method of introducing impurities into a semiconductor. The impurity ions are accelerated electrostatically in a reasonably orthodox accelerator and fired at the semiconductor surface. The depth of penetration is directly determined by the accelerator potentials, and a much wider range of impurity profiles may be produced by varying the accelerator potential as a controlled function of time. Also the impurities tend to go straight into the semiconductor with very little

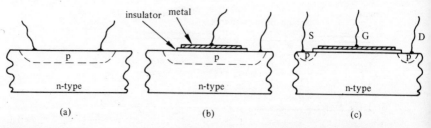

**Figure 3.36.** Various microminiaturised components: (a) resistor; (b) capacitor; (c) MOSFET.

spreading behind any surface masking.  However, one disadvantage of ion implantation is the comparatively small penetration depth.  Potentials of tens of kV are typical, resulting in penetrations of the order of $10^{-6}$ m.

It is clearly essential that different components on the same chip should be electrically isolated from each other, except where deliberately connected.  Figure 3.37 shows a resistor and a transistor, on the same substrate.  Any path from resistor to transistor must pass from a p-region to the n-type substrate and back again into p-type material.  These paths, in other words, will always involve a pair of back-to-back diodes. Consideration of Figure 3.38, for example, shows that, so long as the common point Z is always more positive than both X and Y, then the two diodes will always be reverse-biased and there will be negligible interaction between X and Y.  In practice, this is achieved by bringing to the substrate an external connection, so that its potential may be deliberately made more positive or negative (dependent on whether the substrate is p- or n-type) than any other potential likely to be encountered during the normal operation of the circuit.

Until the advent of integrated circuits, it was necessary for all electronic engineers to be familiar with basic circuit design.  In the future, this will no longer be so important, as a wide range of basic circuits becomes readily available in integrated form.  The engineer will be free to deal with overall systems, without having to consider the actual circuitry involved: 'the architect no longer needs to worry about how the individual bricks are to be made'.

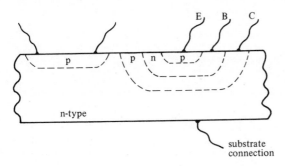

**Figure 3.37.** A resistor and junction transistor on the same substrate.

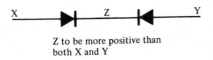

**Figure 3.38.** The isolation of two points connected by back-to-back diodes.

**References**

Bardeen, J., and Brattain, W. H., 1948, *Phys. Rev.,* **70**, 230.

Early, J. M., 1952, *Proc. Inst. Radio Engrs.,* **40**, 1401.

Eimbinden, S., 1964, *Electronics,* **37**, 45.

Emeis, R., Herlet, A., and Spenke, E., 1958, *Proc. Inst. Radio Engrs.,* **46**, 1220.

Fuller, C. S., and Ditzenberger, J. A., 1956, *J. Appl. Phys.,* **25**, 1439.

Grosvalet, J., Motsch, C., and Tribes, R., 1963, *Solid-State Electron.,* **6**, 65.

Handler, P., 1964, *Proc. Inst. Elec. Electron. Engrs.,* **52**, 1444.

Howell, E. K., 1964, *Appl. Note, G.E.C., Rectifier Components Div.*

Jenny, D. A., 1958, *Proc. Inst. Radio Engrs.,* **46**, 959.

Lathrop, J. W., 1964, *Proc. Inst. Elec. Electron. Engrs.,* **52**, 1430.

Lesk, I., and Mathis, V., 1953, *Inst. Radio Engrs. Natl. Conv. Record,* **1**, 2.

Mackintosh, I. M., 1958, *Proc. Inst. Radio Engrs.,* **46**, 1229.

Marinace, J. C., 1960, *IBM J. Res. Develop.,* **4**, 248.

Moll, J. L., 1955, *Proc. Inst. Radio Engrs.,* **43**, 1807.

Shockley, W., 1952, *Proc. Inst. Radio Engrs.,* **40**, 1365.

Shockley, W., Sparks, M., and Teal, G. K., 1951, *Phys. Rev.,* **83**, 151.

Sparkes, J. J., 1960, *Proc. Inst. Radio Engrs.,* **48**, 1696.

Tanenbaum, M., and Thomas, D. E., 1956, *Bell System Tech. J.,* **35**, 1.

Weimer, P. K., Borkan, H., Sadasiv, G., Meray-Horvath, L., and Shallcross, F. V., 1964, *Proc. Inst. Elec. Electron. Engrs.,* **52**, 1479.

# Photodetectors

## 4.1 Photoconductivity

When light is absorbed by a semiconductor, the concentration of free carriers usually rises and this, in turn, raises the electrical conductivity. It is also possible for light to alter the conductivity through a change in the mobility of the charge carriers, rather than their concentration.

In this section we consider the detection of radiation in or near the visible region of the electromagnetic spectrum, reserving the next section for the special problems that arise in the detection of long-wavelength infrared radiation. Quite generally, we require a high sensitivity in the relevant part of the spectrum, a high speed of response, and a low dark conductivity.

Free electron–hole pairs are created in a semiconductor when photons of energy greater than $E_g$ are absorbed. In typical photoconductors it is unusual for these pairs to recombine without the occurrence of certain intervening processes. Recombination takes place after the capture of electrons or holes (or both types of carrier) by centres that arise from impurities and other imperfections, and which have energy levels that lie within the forbidden gap. These centres are of two kinds, namely *traps* and *recombination centres*. When a carrier is captured by a trap, it is more probable that it will be released (by thermal vibration) to that band in which it contributes to the conduction processes, rather than combine with a free carrier of the opposite type. The reverse is true for carriers that are captured by recombination centres. It should be noted that centres can change their role according to whether they are mostly filled with electrons or mostly empty.

If there were no imperfection centres and if the electrons and holes were equally mobile, one would not expect the gain of a photoconductor to exceed unity. The number of electrons flowing through the external circuit would then be no greater than the number of photons absorbed, since the applied field would sweep out all those excited carriers that did not recombine before reaching the electrodes. In practice, however, photoconductive gains that are very much greater than unity are often observed.

In a typical photoconductor, such as cadmium sulphide CdS which has an energy gap of 2·4 eV and is, therefore, intrinsically sensitive to wavelengths that are less than about 0·5 $\mu$m, the excited electrons remain free for a relatively long time, whereas the holes are captured by traps almost immediately. The current flow is thus due almost entirely to the free electrons. Because of the positive charge of the trapped holes, the electrons which flow out at the anode are replaced by others that enter from the cathode, if we assume, of course, that this electrode is a good electron emitter. Current will continue to flow until all the trapped holes have recombined with electrons. If the time for recombination is much longer than the transit time for an electron moving from one electrode to the other, the photoconductive gain becomes much greater than unity.

Consider then, the very simple sequence of events: (i) absorption of photons creates electron-hole pairs, (ii) the holes are immediately trapped, and (iii) the electrons recombine with the holes in the traps with a characteristic relaxation time $\tau$. Events (i) and (ii) are illustrated in Figure 4.1. The density of trapped holes $p_T$ should obey a relation of the form

$$\frac{dp_T}{dt} = -\frac{p_T}{\tau} + g , \tag{4.1}$$

where $g$ is the rate of generation of electron-hole pairs per unit volume, on the assumption of a high trap density.

Now the concentration $n$ of free electrons will be equal to $p_T$, so that the conductivity $ne\mu_n = p_T e\mu_n$. If, then, an electric field $\&$ is applied to the photoconductor, the current density $j$ is equal to $\& p_T e\mu_n$.

It is common to use a photoconductor to detect light that has been modulated at some frequency $f$, the output signal being fed into an amplifying system that is tuned to this frequency. Then, in the steady state, the ratio of the amplitude $I_0$ of the modulated current to the amplitude $g_0$ of the modulated photon absorption is given by

$$\frac{I_0}{g_0} = \frac{\tau \& e\mu_n}{(1 + 4\pi^2 f^2 \tau^2)^{\frac{1}{2}}} , \tag{4.2}$$

this ratio being a measure of the sensitivity of the device [1]. Thus, at low frequencies $f \ll 1/\tau$, the sensitivity is proportional to $\tau\mu_n$, which is to be expected since $\tau \& \mu_n$ is the average number of electron transits before

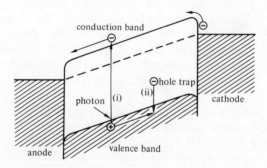

**Figure 4.1.** Excitation of electron-hole pair, hole trapping, and electron transit in an ideal photoconductor.

[1] The specific sensitivity is defined as the product of the photoconductance and the square of the electrode spacing divided by the absorbed light power.

recombination, for a sample of unit length[2]. The sensitivity becomes inversely proportional to the frequency of modulation when $f \gg 1/\tau$.

In the simple model discussed above the response time $\tau_0$ is identical with the recombination time $\tau$, but in real photoconductors it is often found that $\tau_0 \gg \tau$. This is most undesirable, since it is $\tau$ rather than $\tau_0$ that determines the sensitivity according to Equation (4.2). A long response time results from the trapping of electrons which become released only very slowly when the illumination is interrupted (Bube, 1960).

Consider the concentrations, $n$ of electrons in the conduction band, and $n_T$ of trapped electrons per energy interval $kT$ in the region of the Fermi level during excitation (Rose, 1951). Then, in a short time interval $\Delta t$ ($\ll \tau_0$) after the light source is removed, the number of electrons which recombine will be $n \Delta t / \tau$, while $n_T \Delta E / kT$ electrons enter the conduction band from the traps, $\Delta E$ being the amount by which the Fermi level falls in time $\Delta t$. But the overall loss of electrons from the conduction band is $n \Delta t / \tau_0$, which implies that the Fermi level must fall by $kT(\Delta t / \tau_0)$. Thus

$$\frac{n \Delta t}{\tau_0} = \frac{n \Delta t}{\tau} - \frac{n_T \Delta t}{\tau_0}$$

and

$$\tau_0 = \tau \frac{n + n_T}{n} . \tag{4.3}$$

When $n_T \gg n$, which holds when $\tau_0 \gg \tau$,

$$\tau_0 \approx \tau \frac{n_T}{n} . \tag{4.4}$$

It will be seen from Equation (4.1) that, in the steady state,

$$n = g\tau . \tag{4.5}$$

Thus the response time is given by

$$\tau_0 \approx \frac{n_T}{g} . \tag{4.6}$$

Equation (4.6) shows that, for a given concentration of electron traps, the response time increases as the intensity of illumination falls. It has, in fact, been demonstrated by Bube (1965), from this equation, just how difficult it is to achieve a fast response at low light intensities.

[2] In special circumstances it is possible to combine a short response time with a high sensitivity. This desirable state of affairs is achieved if the active centres change their role with the level of illumination. The release time for the trapped holes should be long when light is incident on the photoconductor, but should become short as soon as the light is interrupted and some recombination has taken place. This mode of operation is possible only for a specific level of illumination at any particular temperature.

Suppose that the incident light has a wavelength of $0 \cdot 5$ μm and the number of photons striking the surface of the photoconductor is $4 \times 10^7$ m$^{-2}$ s$^{-1}$ (this is the order of magnitude expected under the sky at night). If it is assumed that these are all absorbed in a depth $d$ metres, the response time must be equal to $n_T d/4 \times 10^7$ s. It is quite unreasonable to expect a density of traps that is less than, say, $10^{18}$ m$^{-3}$, which is the order of the lower limit for the density of imperfections in good quality germanium. Then, if these traps were uniformly distributed over a forbidden gap of 2 eV, the density $n_T$ per interval $kT$ at room temperature would be about $10^{16}$ m$^{-3}$. With the highest conceivable absorption coefficient, the photons might be absorbed within about $10^{-7}$ m of the surface of the sample. Inserting these values in the expression for $\tau_0$, the response time has the value 25 s. These considerations make clear the great importance of purity and perfection of the material for the achievement of fast response times.

## 4.2 Infrared detectors

### 4.2.1 Thermal and photoelectric detectors

There is a rapidly growing interest in far-infrared spectroscopy, and, hence, in equipment for such work. An essential part of this equipment is a range of suitable detectors. The region of wavelengths to be covered extends up to about 1 mm (or 1000 μm); at longer wavelengths than this, the techniques are based on microwave electronics rather than optical principles.

Thermal detectors, such as the thermopile, the bolometer, and the Golay cell (a pneumatic detector) can be used at all wavelengths but suffer from the disadvantage of a relatively long response time. This is because it is necessary to heat the whole of the working substance to obtain a response. Photoconductors, on the other hand, can have a much shorter response time since, when they are used, only the electrons need to become 'heated'. Generally, photoconductors have a restricted bandwidth, and different materials must be used at different wavelengths. However, this can be an advantage in that it allows one to eliminate noise at wavelengths other than that for which the device is designed.

It is interesting to consider the sensitivity that might be expected from ideal thermal and photoconductive detectors. One is usually concerned with the detection of very small amounts of radiant power, and, since bodies at ordinary temperatures emit significant amounts of radiation in the infrared region, the background noise is often comparable with the signal. Thus, the concepts of sensitivity and gain that have been used in discussing photoconductivity at visible wavelengths are not particularly useful here. The quality of an infrared detector is usually expressed in terms of its noise equivalent power $P_N$ or its detectivity $D$.

The noise equivalent power is the flux of infrared radiation that gives a signal-to-noise ratio of unity. The detectivity $D$ is merely the reciprocal of $P_N$. Very often a quantity known as the specific detectivity $D^*$ is employed. $D^*$ is defined as $DA^{\frac{1}{2}}\Delta f^{\frac{1}{2}}$, where $A$ is the area of the device and $\Delta f$ is the frequency bandwidth specified for the noise equivalent power ($f$ being the frequency of modulation of the signal). $D^*$ is equal to the detectivity for a cell of unit area, the output of which is fed into an amplifier of unit bandwidth, and is useful in comparing different detectors. It must, however, be used with caution since not all detectors have the same area and bandwidth dependence of the detectivity (Putley, 1965).

In considering the theoretical performance of an ideal detector, it is supposed that a fluctuation of the background radiation is superimposed upon the modulated signal. It can be shown (Lewis, 1947) that the mean square fluctuation in the power received by an area $A$ from surroundings at temperature $T$ (with a field of view of $2\pi$ steradians) is

$$\overline{\Delta w^2} = 8\epsilon\sigma kT^2 A\Delta f,\tag{4.7}$$

where $\epsilon$ is the emissivity of the detector and $\sigma$ is Stefan's constant.

Thus, for an ideal thermal detector with $\epsilon = 1$, the noise equivalent power is

$$P_N = (2\overline{\Delta w^2})^{\frac{1}{2}} = 3\cdot56 \times 10^{-17}\, T^{\frac{5}{2}}A^{\frac{1}{2}}\Delta f^{\frac{1}{2}}\ \text{watts}.\tag{4.8}$$

$P_N$, $D$, and $D^*$ are all independent of wavelength for such a detector.

For the ideal thermal detector the output power is equal to the input power at all wavelengths, but this is not true for the ideal *photoconductive*

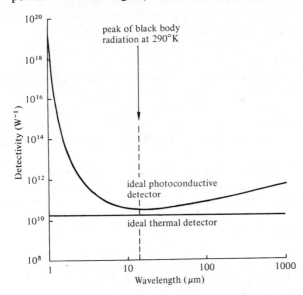

**Figure 4.2.** Dependence of detectivity on signal wavelength for ideal thermal and photoconductive detectors. The background radiation is supposed to come from a black-body at 290°K (Putley, 1965).

detector. If the signal wavelength is $\lambda_s$, the activation energy for the carriers should be made equal to $hc/\lambda_s$, and no carriers can then be excited for longer wavelengths. In other words, noise at these longer wavelengths is eliminated. Also, if we assume that the *quantum* efficiency is unity for all wavelengths $\lambda$, less than $\lambda_s$, the *power* efficiency in this region is $\lambda/\lambda_s$. Thus, the noise for $\lambda \ll \lambda_s$ is much less well detected by the ideal photo-conductor than by the ideal thermal detector. The variation of detectivity with wavelength for the ideal detectors is shown in Figure 4.2 for background radiation from a black body at 290°K. The minimum in the detectivity of the photoconductor occurs at more or less the same wavelength as the peak of the black-body radiation.

In this chapter we shall confine our attention to photoelectric detectors, but one or two types of thermal detector will be mentioned in other chapters.

### 4.2.2 Intrinsic and extrinsic detectors

For the detection of wavelengths up to the order of 10 $\mu$m one makes use of *intrinsic* photoconductivity, that is the excitation of electron–hole pairs across the energy gap. There is, however, one important difference between the uses of intrinsic photoconductivity for visible and infrared radiation. The energy gap of a photoconductor that is sensitive to visible light is large enough for the dark conductivity to be made very small, but the decrease in the gap width that is necessary for the detection of infrared rays means that the dark conductivity at room temperature must rise. It is, therefore, good practice to cool all infrared photoconductors, and particularly those that are used at the longer wavelengths.

The lead chalcogenides PbS, PbSe, and PbTe have been widely used as intrinsic photoconductors up to wavelengths of about 5 $\mu$m. Table 4.1 shows the energy gaps and the corresponding maximum wavelengths $\lambda_s$ at which radiation can be detected by the three compounds. It will be seen that, by cooling it to liquid nitrogen temperature, a lead selenide detector can be used out to 6·5 $\mu$m. For most semiconductors the energy gap falls as the temperature rises, but the lead chalcogenides are an exception to this rule.

In order to gain the full advantage from the cooling of an infrared photoconductor it is important that it should be thoroughly pure. If it is impure, the extrinsic carriers arising from the donor or acceptor impurities

**Table 4.1.** Energy gaps and limits of detection for the lead chalcogenides.

| Compound | 300°K | | 77°K | |
|---|---|---|---|---|
| | $E_g$ (eV) | $\lambda_s$ ($\mu$m) | $E_g$ (eV) | $\lambda_s$ ($\mu$m) |
| PbS | 0·41 | 3·0 | 0·32 | 3·8 |
| PbSe | 0·29 | 4·3 | 0·19 | 6·5 |
| PbTe | 0·32 | 3·9 | 0·22 | 5·7 |

will increase the dark conductivity. Also, the movement of the Fermi level into the conduction or valence band, when the extrinsic carrier concentration is large, leads to a shift of the absorption edge and the limit of detectivity moves to smaller wavelengths (the Burstein–Moss effect).

A disadvantage of the lead salts is that it is difficult to prepare them with even approximately the stoichiometric composition. Departures from stoichiometry, of course, lead to high carrier concentrations, just like additions of foreign atoms. On the other hand, the III–V compounds can be produced in almost stoichiometric form with very small carrier concentrations. Both InSb and InAs, with energy gaps of $0 \cdot 18$ eV and $0 \cdot 35$ eV at $300°$K, can be used as infrared photoconductors; the corresponding wavelength limits are $7 \cdot 3$ $\mu$m and $3 \cdot 5$ $\mu$m. The detectivity of an indium antimonide photoconductor at room temperature falls far short of the ideal value, owing to Johnson noise. However, at $77°$K the detectivity lies very close to that of the ideal photoconductor. This increase of detectivity is gained at the expense of a fall in the wavelength limit to about $5 \cdot 5$ $\mu$m, owing to an increase in the energy gap.

In order to extend the range of intrinsic photoconductors further into the infrared region, it is necessary to find semiconductors of still smaller energy gap. In principle, it is now possible to achieve any gap width down to zero. This can be done, for example, by alloying a semi-metal (which has overlapping conduction and valence bands) with a semiconductor. Thus the semi-metal HgTe forms a continuous range of solid solutions with CdTe (Lawson *et al.*, 1959), allowing energy gaps between zero and $1 \cdot 45$ eV to be obtained.

Of special interest is the system of alloys between PbTe and SnTe, since both these compounds are semiconductors. As shown in Figure 4.3, the

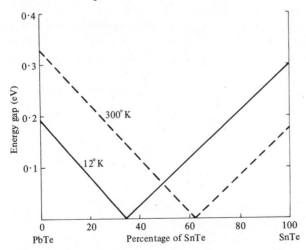

**Figure 4.3.** Energy gap plotted against composition for the PbTe–SnTe pseudo-binary system (Dimmock *et al.*, 1966).

energy gap falls both on adding PbTe to SnTe and on adding SnTe to PbTe; a zero gap width is observed at some intermediate composition. The energy gap of PbTe rises with increasing temperature, while that of SnTe falls so that the composition for zero gap changes with temperature (Dimmock *et al.*, 1966). The reason for the observed behaviour is an inversion of the valence and conduction bands that occurs as the composition passes through the zero gap value. It has been found that it is actually easier to achieve stoichiometry between the electropositive and electronegative elements in (Pb–Sn)Te than in the simple compound PbTe.

In spite of the increasing availability of narrow-gap semiconductors and strenuous efforts to purify them, there is a practical limit to the wavelength for which intrinsic photoconductors can be used. Beyond this limit, it is necessary to exploit the excitation of electrons or holes, by photons, from impurity levels that lie within the forbidden gap. In other words, *extrinsic* photoconductors must be employed. The disadvantage of an extrinsic photoconductor is that it absorbs photons much less strongly than does the corresponding narrow-gap intrinsic material. It is, of course, essential to cool these extrinsic photoconductors, sometimes to liquid helium temperature, to prevent thermal excitation of the carriers from the impurity states.

Germanium is commonly employed as an extrinsic photoconductor since it is easy to obtain samples of the requisite purity, and the behaviour of a large number of different impurities is well known. The hydrogenic

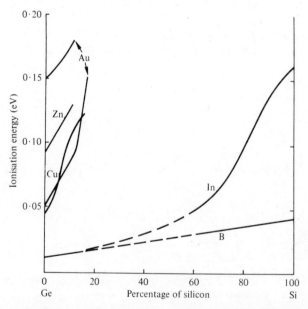

**Figure 4.4.** Impurity activation energies for various impurities in Ge–Si alloys (Brown and Kimmitt, 1965).

model for an impurity centre gives a value for the activation energy
$E_{imp} = 13 \cdot 6(m*/m)/\epsilon_r^2$ eV, where $m*$ is the effective mass of the carriers
and $\epsilon_r$ is the relative permittivity. For electrons in germanium
$m*/m \approx 0 \cdot 25$ and $\epsilon_r \approx 16$, indicating that the impurity activation energy
should be about $0 \cdot 013$ eV. In practice, the elements of group III and V
of the Periodic Table do behave approximately as predicted by the
hydrogenic model when added to germanium, and have activation energies
of the order of $0 \cdot 01$ eV. Impurities from other groups have a wide range
of somewhat larger activation energies, allowing a reasonably good
detectivity to be achieved at wavelengths from about 10 $\mu$m up to the
order of 100 $\mu$m. In fact, when germanium is alloyed with silicon, the
activation energy $E_{imp}$ for any particular impurity is increased, and, by
choice of a suitable Ge-Si alloy, any specific value of $E_{imp}$ between wide
limits can be obtained, as shown in Figure 4.4 (Shultz and Morton, 1965).

### 4.2.3 Free-electron photoconductivity

It might seem that extrinsic photoconductivity, using impurity centres in a
semiconductor such as InSb, which has a very small effective mass
$(m*/m \sim 1/100)$ could be used in the detection of radiation having a
wavelength in excess of 100 $\mu$m. This turns out to be impossible at the
present time because, if the impurity concentration exceeds about $10^{13}$ cm$^{-3}$,
the impurity levels broaden into bands that overlap the main bands. Even
if pure enough material were available, it would have a very small
absorption coefficient. Thus, for $\lambda > 100$ $\mu$m, one uses a free-carrier
photoconductive effect.

In a free-electron photoconductor it is the energy, rather than the
concentration of carriers, that is changed by the radiation. Since the
semiconductor is held at liquid helium temperature, the thermal energy of
the carriers can be small. Thus, the average energy can be increased
profoundly by radiation of even long wavelength, and, if the relaxation
time is energy-dependent, the mobility is altered. In effect the sample acts
as a bolometer, but it is the electron gas rather than the lattice that
becomes hot. Consequently the response time is much shorter than that
of a conventional bolometer.

### 4.2.4 Effects of a magnetic field

In the near-infrared region, the intrinsic photoconductive detectors are
sometimes replaced by detectors that make use of the photo-electromagnetic
effect. The origin of this effect is shown in Figure 4.5. Electron-hole
pairs that originate near the illuminated surface drift into the bulk of the
sample because of their concentration gradient. This drift is given a
transverse component, in the opposite sense for electrons and holes, by
the magnetic field. Thus, an e.m.f. appears between the side faces. The
photo-electromagnetic detector has a very short response time (Zitter,
1964) but presents practical difficulties because of its low electrical
resistance.

The most important use of a magnetic field has been in the improvement of the performance of detectors of far-infrared radiation. The material used in the free-electron photoconductive detector, described above, is usually InSb, since this compound displays very marked hot electron effects (i.e. the mobility changes, through the energy dependence of the relaxation time, at quite low electric fields). However, there are difficulties due to the rather small resistivity ($\sim 10^{-1}$ $\Omega$ m) of even the best available InSb. The resistivity can be increased at about $2°$K by three orders of magnitude using a modest magnetic field of $0 \cdot 6$-$0 \cdot 7$ T (Putley, 1965).

Much larger magnetic fields than this can easily be reached using super-

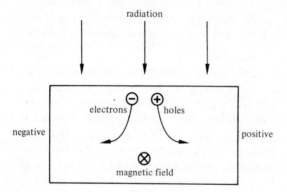

**Figure 4.5.** Origin of the photo-electromagnetic effect. The magnetic field is normal to the section shown.

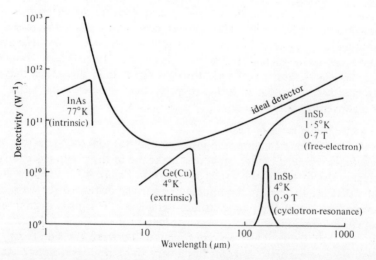

**Figure 4.6.** Detectivity of typical photoconductive detectors in the infrared region (after Putley, 1965).

conducting solenoids (see Chapter 8). If a field in excess of about 1 T is applied to an InSb detector, its performance can be improved, though its mode of operation is then completely different. It becomes sensitive over a narrow band of wavelengths centred at the cyclotron resonance value (the angular frequency of cyclotron resonance is given by $\omega = eB/m^*$, where $B$ is the magnetic field). This increased sensitivity is associated with the strong absorption of radiation when the cyclotron resonance condition is satisfied. Brown and Kimmitt (1965) have found that, as the magnetic field is changed from $1 \cdot 4$ T to $7 \cdot 6$ T, the wavelength for maximum sensitivity moves from 100 $\mu$m to 26 $\mu$m. The tunable cyclotron-resonance detector offers the possibility of a really high detectivity since it can be made insensitive except in the region of the signal wavelength.

Figure 4.6 shows how some of the detectors that have been discussed compare with the ideal photoconductor at various wavelengths (Putley, 1965). It should be noted that the junction devices that will be treated in the next section can also be used at infrared wavelengths, provided that they are manufactured from suitable materials.

## 4.3 Junction photodetectors

### 4.3.1 Photodiodes

A photoconductor must be supplied with current from an external power source. There is, however, another type of device, the photovoltaic cell, that itself acts as a source of e.m.f. when it is illuminated. In a typical photovoltaic cell, light is absorbed near a p-n junction in a semiconducting crystal; the electrons and holes that have been excited by the photons then move under the influence of the internal field that exists across the depletion layer. Although light meters based on the photovoltaic effect have been in use for many years, it is only since the techniques of transistor production have been applied to photocells made from silicon and other well-behaved semiconductors that reasonably high efficiencies have been achieved.

The origin of the photovoltaic effect at a p-n junction is shown in Figure 4.7. (The Fermi level is drawn as a horizontal line, which implies that the cell has been short-circuited.) Electrons and holes, that are excited by light absorbed in the depletion layer, are swept to the n and p regions respectively. Photo-excited minority carriers that lie outside depletion regions are not immediately subject to a high field. They can, however, diffuse towards the junction and, once they reach the depletion layer, they drift rapidly across. This process, of course, depends on the minority carrier lifetime being large enough for the excited electrons or holes to avoid recombination while they diffuse in the region of almost zero field. The active part of the device comprises the depletion layer and the regions about a diffusion length wide on either side. Thus, one can ensure that nearly all the excited carriers cross the junction by making the depletion layer wide or by increasing the minority carrier lifetimes. However, if a short response time is required, the carrier lifetimes should be low, and the carriers should drift in an electric field rather than diffuse. The desirable wide depletion layer results from a low impurity concentration on one side of the junction or the other.

Let us consider the current-voltage characteristics of a junction diode in the dark and when illuminated. The behaviour is shown in Figure 4.8. If

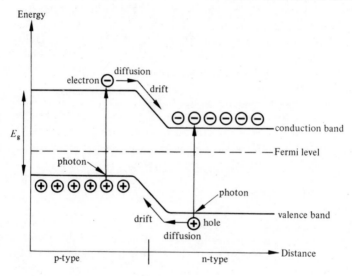

**Figure 4.7.** Photovoltaic effect at a p-n junction.

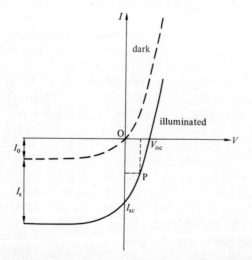

**Figure 4.8.** Current-voltage characteristic for a junction diode when illuminated.

the series resistance of the diode is ignored the dark current is given by

$$I = I_0 \left[ \exp\left(\frac{eV}{kT}\right) - 1 \right].$$
(4.9)

It is supposed that, upon illumination, there is an additional photocurrent $-I_s$ (it has been given a negative sign since it flows in the opposite direction to the forward current of the diode). If $P$ is the flux of photons and $\eta$ is the quantum efficiency, $I_s$ must be equal to $\eta eP$. The current on illumination is

$$I = I_0 \left[ \exp\left(\frac{eV}{kT}\right) - 1 \right] - I_s$$
(4.10)

and the voltage is given by

$$V = \frac{kT}{e} \ln\left(\frac{I + I_s + I_0}{I_0}\right).$$
(4.11)

The open-circuit voltage $V_{oc}$ is equal to $(kT/e)\ln[(I_s+I_0)/I_0]$ and the short-circuit current $I_{sc}$ is equal to $-I_s$. The power output reaches a maximum when the area of the rectangle OP in Figure 4.8 is largest; it is approximately equal to $0 \cdot 75 V_{oc}/I_{sc}$. This is the condition that should be obeyed when the diode is used as a solar cell, i.e. a convertor of solar energy into useful electrical energy. It must, of course, operate within the bottom right-hand quadrant of Figure 4.8 if there is to be no external electrical power supply.

If the cell is to be used as a sensitive radiation detector, it is advantageous to supply a reverse-bias voltage, so that the bottom left-hand quadrant of the current–voltage characteristic becomes relevant. At a high reverse voltage the current is $-(I_s+I_0)$ which is approximately equal to $-I_s$ if the saturation reverse current $I_0$ in the dark is very small. On illumination, the change in the power delivered to a high-resistance load is $-I_s V$ and this can be very much greater than the power output at zero bias. Reverse-bias increases the width of the depletion region as well as the field within that region. This is particularly important in improving the speed of response at wavelengths that are close to the absorption edge, especially for a semiconductor such as silicon that does not have a very sharp edge. Silicon has an indirect energy gap and, although the absorption edge lies at about $1 \cdot 1$ $\mu$m, the absorption coefficient has risen to only about $10^4$ $m^{-1}$ at a wavelength of $1 \cdot 0$ $\mu$m. A very wide depletion layer is needed, if the minority carriers are to be generated in a high-field region, when the photons are only weakly absorbed.

As it has been described so far, the photodiode differs from the photo-conductor in that it does not display current gain. Current gain can be achieved in a diode detector if the reverse field is large enough to cause impact ionisation (Johnson, 1965). The gain then rises rapidly owing to an avalanche process. If a photodiode is to be used in the avalanche mode, it is very important to eliminate so-called microplasmas, i.e. regions in which the breakdown field is much less than elsewhere. Surface breakdown can be eliminated by a guard-ring technique and the appearance of microplasmas in the bulk is avoided by making the active area small. Avalanche photodiodes compare favourably with fast photoconductive detectors in that high current gain can be achieved even when the recombination rate is high and, thus, the response time is low. There are, however, very few semiconductors in which suitable p–n junctions can be produced.

The solar cell is a particularly interesting application of the photovoltaic effect (Chapin *et al.*, 1954). It is, of course, rather more difficult to select the best material for converting the wide spectrum of solar radiation into electrical energy than for detecting light of a specific wavelength. The power from the Sun per unit bandwidth has its highest value at a wavelength of about $0\cdot5$ $\mu$m, but it still has about half this value at a wavelength of 1 $\mu$m. Thus, a diode made from a semiconductor with a band gap of about 2 eV is very effective in converting the solar energy at the peak of its wavelength distribution into electricity, but it wastes all the energy in the infrared region. On the other hand, a diode that utilises a semi-conductor with a gap of only $0\cdot5$ eV would certainly convert nearly all the photons arriving from the Sun into electrical energy, but much of the energy in the visible and near-infrared regions below 1 $\mu$m would be lost. This is because only a fraction of the energy of each photon could be absorbed by inter-band edge-to-edge transitions. Clearly the band gap must be chosen as a compromise between two conflicting requirements. It must be remembered that the saturation reverse current $I_0$ in the dark depends to some extent on the energy gap, and this must also be taken into account in the selection of materials.

It has been calculated (Rappaport and Wysocki, 1965) that solar cells made from the best of the known semiconductors, GaAs, should have a maximum efficiency of 24% but practical cells have an efficiency of 11%. Silicon has a theoretical maximum efficiency of only 20% but, since its technology is more advanced, it has yielded solar cells with an efficiency, 14%, that is substantially higher than that of GaAs cells. The construction of a silicon solar cell is indicated in Figure 4.9. The junction is situated about 1 $\mu$m below the receiving surface, this distance being chosen so that most of the excited electrons and holes can contribute to the output. The photons in the visible region are absorbed very close to the surface, whereas the infrared photons penetrate much more deeply. An interference coating of the appropriate thickness and refraction index is used to minimise reflection losses. Great care is taken to make the series resistance of the cell as low as possible by ensuring that the contacts are ohmic and sometimes by using a grid structure for the top electrode.

Cadmium sulphide (energy gap $2 \cdot 4$ eV) would not seem to be the ideal solar cell material and, in fact, has a theoretical maximum efficiency of only 16%. The experimentally observed efficiency of 7% compares rather unfavourably with that of the best silicon cells, but it has been found that CdS cells can be made in the form of large-area plates from polycrystalline material. Such cells can be produced at a much lower cost per unit area than silicon cells.

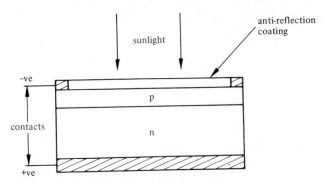

**Figure 4.9.** Construction of a silicon solar cell.

### 4.3.2 Phototransistors
The phototransistor, like the avalanche photodiode, is a detector that exhibits current gain. It can, in fact, be regarded as the combination of a simple photodiode and a transistor. A p-n-p phototransistor is illustrated in Figure 4.10. Light is absorbed in the base region which is not connected to the external circuit. A reverse-bias exists at the base-collector junction due to the voltage applied between the emitter and the collector. Little current flows when the base is not illuminated.

Upon illumination, holes that are excited in the base region diffuse out, leaving behind an overall negative charge. This charge forward-biases the emitter-base junction so that holes are injected into the base and move across into the collector. A current of holes from the emitter to the collector continues until the negative charge of the excess electrons in the base is neutralised by recombination.

The phototransistor is similar to the photoconductor in that its gain and response time rise together. A high gain depends on a long recombination time for the excess electrons that are held in the base region.

### 4.4 Electrophotography
In the various electrophotographic processes a latent image is formed, as in ordinary photography, by exposing a layer of sensitive material to light. This latent image takes the form of a charge pattern on an electrical insulator; it can be developed by applying a fine powder to the surface.

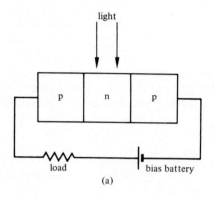

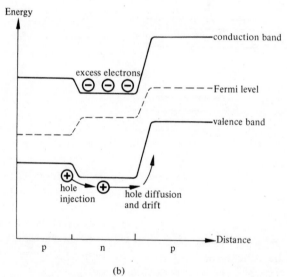

**Figure 4.10.** The phototransistor: (a) schematic arrangement; (b) energy diagram.

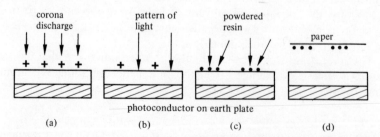

**Figure 4.11.** Electrophotography—the Xerography process: (a) the photoconductor is charged and (b) discharged in selected regions by illumination, (c) fine resin particles are attracted to the charged parts of the photoconductor and (d) the particles are transferred onto paper.

The powder distributes itself according to the charge pattern, through electrostatic attraction.

Perhaps the best known of the techniques is Xerography (Schaffert and Oughton, 1948). Here the layer consists of a thin coating of amorphous selenium, which is a photoconductor having a dark conductivity of only about $10^{-14}$ $\Omega^{-1}$ $m^{-1}$, on a metal surface. The surface of the layer is first charged all over by means of a corona discharge. The pattern of light to be recorded is then allowed to fall on the layer. The illuminated regions become conducting and permit their surface charge to leak away, while the dark regions remain charged. A cloud of finely powdered pigmented resin is then applied to the surface and the particles cling to the charged regions. The powder pattern is transferred to a sheet of paper and bonded to it by heating. The various steps are shown in Figure 4.11.

In another process, known as Electrofax, a sensitised paper replaces the photoconductive plate (Young and Greig, 1954). The paper is coated with a layer of the white compound zinc oxide, in the form of fine grains dispersed in a resin. This photoconductive layer is charged and parts of it are discharged, as described above, to form the required charge pattern. A black dust is applied and fused to the regions where it collects. An advantage of this process is that there is no loss of contrast arising from the transfer of the powder from one surface to another.

**References**
Brown, M. A. C. S., and Kimmitt, M. F., 1965, *Infrared Phys.,* **5**, 93.
Bube, R. H., 1960, *Photoconductivity of Solids* (Wiley, New York).
Bube, R. H., 1965, *Photoelectric Materials and Devices,* Ed. S. Larach (Van Nostrand, Princeton), p.100.
Chapin, D. M., Fuller, C. S., and Pearson, G. L., 1954, *J. Appl. Phys.,* **25**, 676.
Dimmock, J. O., Melngailis, I., and Strauss, A. J., 1966, *Phys. Rev. Letters,* **16**, 1193.
Johnson, K. M., 1965, *Inst. Elec. Electron. Engrs., Trans. Electron Devices,* **ED-12**, 55.
Lawson, W. D., Nielsen, S., Putley, E. H., and Young, A. S., 1959, *J. Phys. Chem. Solids,* **9**, 325.
Lewis, W. B., 1947, *Proc. Phys. Soc. (London),* **59**, 34.
Putley, E. H., 1966, *J. Sci. Instr.,* **43**, 857.
Rappaport, P., and Wysocki, J. J., 1965, *Photoelectric Materials and Devices,* Ed. S. Larach (Van Nostrand, Princeton), p.239.
Rose, A., 1951, *RCA Rev.,* **12**, 362.
Schaffert, R. M., and Oughton, C. D., 1948, *J. Opt. Soc. Am.,* **38**, 991.
Schultz, M. L., and Morton, G. A., 1965, *Photoelectric Materials and Devices,* Ed. S. Larach (Van Nostrand, Princeton), p.140.
Young, C. J., and Greig, H. G., 1954, *RCA Rev.,* **15**, 469.
Zitter, R. N., 1964, *Rev. Sci. Instr.,* **35**, 594.

# Lamps and lasers

## 5.1 Emission of light from solids

The familiar tungsten filament lamp is, of course, a solid-state light source. It suffers, however, from several disadvantages, namely a low luminous efficiency and a limited life (these two qualities cannot be regarded as independent since one can be improved at the expense of the other) and a rather ill-defined and unstable source position. It may, in fact, be possible to improve the performance of the incandescent lamp by replacing the tungsten filament by a thin semiconducting plate (Fok, 1962).

Semiconductors have a high absorption coefficient for photons of energy greater than the width of the forbidden band, but the absorption coefficient can be much smaller for wavelengths beyond the main absorption edge. Thus, in principle, it might be possible to find a refractory semiconducting material with an energy gap of about 2 eV, which would have a high absorption coefficient for visible light and a low absorption coefficient for infrared radiation. Now, if the absorptance of a sample is high in a given spectral region, its emittance should also be high (provided that it is not too good a reflector of light); similarly, if its absorptance is low, it will be a poor emitter of radiation. Hence, if the semiconductor mentioned above were stable when heated to a temperature of the order of 2000°K (at which temperature there is a reasonable amount of black-body radiation in the visible part of the spectrum), one might be able to make a selective incandescent emitter of visible radiation. Its efficiency should be much greater than that of the tungsten lamp.

The greatest problem in trying to realise this efficient incandescent lamp probably arises from the increase of absorption coefficient beyond the absorption edge, as the temperature rises, due to the free carriers which have been excited across the energy gap. Thus, for a sample to be more or less transparent at a high temperature, in the infrared region of the spectrum, it is necessary for it to be very thin, probably no more than two or three micrometres thick. At least, this seems to be so for the most promising semiconducting material, silicon carbide (Kauer, 1965). SiC can be heated to a reasonably high temperature without decomposition in a suitable atmosphere, and, although (in its hexagonal form) its energy gap at room temperature is nearer 3 eV than 2 eV, the gap should approach 2 eV at elevated temperatures. It may thus not be out of the question to produce very thin plates of such a material as SiC, but it must be admitted that the prospects for a selective incandescent emitter are not too good at the moment. Thus, for the rest of this chapter we shall be concerned with cold light sources, that is with luminescence.

In discussing luminescence, we shall be concerned with the absorption or emission of light when electrons undergo transitions in a solid from one energy level to another. The absorption of a photon of angular frequency

$\omega$ is accompanied by an electron transition from a state of energy $E_1$ to a state at a higher energy $E_2$, where $\hbar\omega = E_2 - E_1$. When a flux $P$ of photons of the appropriate frequency is incident upon a material, the rate of electronic transitions $n(1 \rightarrow 2)$ per unit volume is given by

$$n(1 \rightarrow 2) = N_1 B_{12} P , \qquad (5.1)$$

where $N_1$ is the concentration of electrons in the state 1 and $B_{12}$ is the transition probability for the absorption process.

Emission of light is accompanied by electron transitions from the higher to the lower energy band. We must distinguish between two types of emission process. First, there is the spontaneous decay of any excess of electrons in the upper state, above the density corresponding to thermal equilibrium. This is the process by which the thermal equilibrium situation is reached when the illumination of a sample is interrupted, if we ignore non-radiative transitions. If the spontaneous transition probability is $A$, the rate of spontaneous transitions is

$$n_{\text{spontaneous}}(2 \rightarrow 1) = N_2 A , \qquad (5.2)$$

where $N_2$ is the concentration of electrons in the upper state 2. Secondly, so long as light is incident on the material, it will induce electronic transitions in the direction $2 \rightarrow 1$ as well as in the direction $1 \rightarrow 2$. If this process (known as stimulated emission) has a transition probability $B_{21}$, the rate of stimulated transitions is

$$n_{\text{stimulated}}(2 \rightarrow 1) = N_2 B_{21} P , \qquad (5.3)$$

$P$ again being the photon flux. It is important to note that, in stimulated emission, the emergent photon has the same frequency as, and is in phase with, the incident photon. This implies that there is amplification of the incident radiation, just as the absorption process implies attenuation.

We must regard absorption itself as a stimulated process and, not surprisingly, the probabilities $B_{12}$ and $B_{21}$ are equal (we set them both equal to $B$). Whether or not a given event will lead to emission or absorption will depend on the phase of the incident photon. At equilibrium, for a given photon flux,

$$(N_1 - N_2)BP = N_2 A , \qquad (5.4)$$

since the number of upward transitions $n(1 \rightarrow 2)$ must be equal to the total number of downward transitions $n(2 \rightarrow 1)$.

Stimulated emission has, of course, been of great importance in recent years with the advent of the maser and laser[1]. In this connection the ratio of the coefficients $B$ and $A$ (known as the Einstein coefficients) is of great significance. The maser is of great importance as a very low-noise amplifier and $B/A$ is a measure of the signal-to-noise ratio for such a

[1] Acronyms for *microwave* and *light amplification by stimulated emission of radiation*.

system. By considering the radiant energy density as a function of frequency from the Planck radiation law it can be shown that $B/A \propto \lambda^3$, where $\lambda$ is the wavelength (Thorp, 1967). Thus, low-noise amplification is easier for microwave radiation than for light, while the laser is far more important as a highly coherent light source rather than as a low-noise amplifier.

Before we consider ordinary luminescent processes, it is worth discussing the conditions that are necessary for laser operation, since, as we shall see later, certain luminescent systems become lasers when operated under the appropriate conditions. If there is to be overall amplification of an incident signal, it is necessary that $n_{stimulated}(2 \to 1)$ should exceed $n(1 \to 2)$. Equations (5.1) and (5.3) show that this can only occur if $N_2 > N_1$, a situation that is at variance with the condition of thermal equilibrium at any temperature. Amplification, then, requires so-called population inversion with a certain high-energy state more densely populated than one of a lower energy; the process for achieving population inversion is known as pumping. In its simplest form, a laser produces highly coherent (that is strongly monochromatic) light by amplifying the radiation of a specific wavelength that originates from a spontaneous transition; no externally applied signal is needed. The possible wavelengths (within a narrow band that is determined by the energy levels in the laser material) are fixed by the boundary conditions, the laser having the form of a Fabry–Perot resonator. Specific laser systems will be discussed later.

Luminescence of any kind will occur only if the situation of thermal equilibrium is disturbed by increasing the population of electrons in some higher-energy state at the expense of some lower-energy state (though not necessarily actually inverting the population). There are various techniques for exciting luminescence including some (e.g. chemiluminescence—the excitation of luminescence using chemical energy) that do not concern us here.

*Electroluminescence*, that is the production of light by passing an electric current through a material, is particularly attractive, and we shall devote most of this chapter to the study of this phenomenon in semi-conductors. In some semiconductors it has been found very difficult to produce the conditions that are necessary for electroluminescence; it has, however, been found possible, in some of these cases, to observe *cathodoluminescence*, that is the excitation of luminescence using an electron beam.

*Photoluminescence*, the production of light of one wavelength on illumination of the material with light of a shorter wavelength, is, of course, of great importance in fluorescent lamps. The ultraviolet light that is emitted by mercury vapour in an electric discharge is converted into visible light by means of special phosphors. Here we shall be more interested in photoluminescence as the basis of solid-state lasers.

## 5.2 Junction electroluminescence and semiconductor lasers

The simplest form of semiconductor lamp consists essentially of a forward-biased junction diode. The energy band diagram, Figure 5.1, shows that electrons are injected into the p-type region while holes enter the n-type region. The injected minority carriers ultimately recombine with majority carriers of the opposite type. Ideally the recombination processes should give rise to electromagnetic radiation, but, in practice, the radiative processes are accompanied by non-radiative transitions in which the energy of the injected carriers is passed on to the thermal vibrations. The radiative processes can consist of band-to-band transitions or can involve impurity centres. It is also possible for radiation to be emitted as a result of transitions within a single band (so-called deceleration emission).

The internal quantum efficiency of the device depends on the ratio of radiative to non-radiative processes, but the overall efficiency also depends on other factors. Thus, the ohmic losses, due to the electrical resistance of the diode, should be as small as possible. Also, it is important that as much of the emitted light as possible leaves the diode instead of being absorbed or internally reflected.

The energy band structure of a semiconductor is the most important factor in determining whether or not it will be a useful electroluminescent material. The width of the energy gap will govern the wavelength of the emitted light, while the probability of radiative transitions depends largely on whether the gap is 'direct' or 'indirect'. The difference between direct and indirect gaps is shown in Figure 5.2. For a direct-gap semiconductor (Figure 5.2a) the valence band maximum and conduction band minimum have the same value for the wavevector (usually but not always at k = 0); since the wavevector of a photon is almost negligible, it is then simple for an electron and hole to recombine with the emission of light while at the same time conserving both energy and wavevector.

Indirect transitions (shown in Figure 5.2b) occur when the extrema of the valence band and conduction bands occur at different wavevector

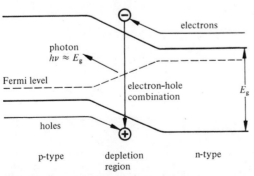

**Figure 5.1.** Carrier injection and recombination in a forward-biased junction diode.

values. In this case, the recombination process must be accompanied by the emission or absorption of a phonon if the conservation laws are to be satisfied. Thus the photon energy is given by

$$\hbar\omega = E_C - E_V \pm E_{ph} , \qquad (5.5)$$

where $E_V$ and $E_C$ are energies at (or near) the edges of the valence and conduction bands and $E_{ph}$ is the phonon energy. It should be noted that $E_C - E_V \gg E_{ph}$ in all cases. For the conservation of wavevector

$$\mathbf{k}_C - \mathbf{k}_V = \pm\mathbf{q} , \qquad (5.6)$$

where $\mathbf{q}$ is the wavevector of the phonon. The indirect process is radiative but its transition probability is much lower than that of a direct process. Consequently, the competing non-radiative processes are relatively more frequent in an indirect-gap semiconductor than in a direct-gap material. Impurity states having energies close to those of the band edges are closely associated with the states in the main bands. Thus, one can still think in terms of direct or indirect transitions when such shallow impurity states are involved. However, this is no longer true for deep impurity states, and it is possible, in principle, to obtain strong radiative recombination in indirect-gap semiconductors with their aid.

At present, junction technology has reached an advanced stage only for silicon, germanium, and certain of the III-V compounds. Silicon and germanium have indirect energy gaps and are unsuitable for use in light-emitting diodes. However, several of the III-V compounds have direct energy gaps. The most useful of these is GaAs with an energy gap of $1.35$ eV, giving electroluminescence at about $0.9$ $\mu$m. This is, of course, in the infrared region of the spectrum but for many applications this does

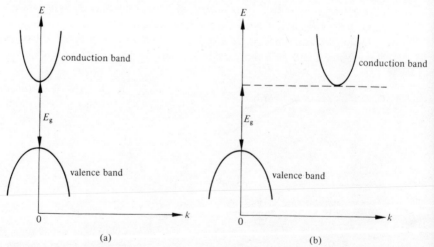

**Figure 5.2.** $(E, k)$ diagrams for (a) a direct-gap semiconductor and (b) an indirect-gap semiconductor.

not matter. Thus, some systems make use of the fact that silicon photo-cells have a high efficiency at wavelengths corresponding to the light output of GaAs diodes. The compound GaP has an energy gap of $2 \cdot 25$ eV which is certainly large enough for the emission of visible radiation, but unfortunately the gap is of the indirect type. Nevertheless, by suppressing the non-radiative processes, worthwhile GaP lamps have been made. GaP diodes with overall efficiencies of the order of 1% for deep-red light can be produced (Logan *et al.*, 1965). These diodes make use of transitions involving deep impurity levels; emission due to transitions between levels near to the band edges leads to an efficiency that is an order of magnitude lower, but the emitted light is then green and, since the eye is most sensitive to this colour, the luminous efficiency is not much different (Pilkuhn and Foster, 1966).

It is possible to raise the energy gap of GaAs, without it changing from the direct to the indirect type, by forming solid solutions of that compound with either GaP or AlAs. Of course, since AlAs, like GaP, is an indirect-gap semiconductor, it is essential that the concentration of GaAs is not made too small. It has been reported that direct transitions with a wavelength as low as $0 \cdot 62$ $\mu$m, which is in the red region of the visible spectrum, can be observed in a GaAs–AlAs alloy (Black and Ku, 1966).

There are other materials that can be used in electroluminescent diodes besides those mentioned above. For example, SiC lamps having a range of colours from red to blue (Brander, 1965) have been prepared but their efficiencies are not at all impressive, even compared with lamps made from GaP. It is expected that efficient junction electroluminescence in the visible region will eventually be obtained, probably using a direct-gap semiconductor, so it is rather more useful to discuss the construction of the efficient GaAs infrared lamp rather than the inefficient GaP visible lamp.

It is important to remember that the diffusion length for minority carriers in a direct-gap semiconductor must always be very small. Now one of the conditions for a low forward resistance of a diode is that the diffusion length should be greater than the width of the depletion layer, since the current in this layer is carried by injected carriers rather than equilibrium carriers. Thus one wants a narrow depletion layer, which is achieved by forming an abrupt p–n junction. The resistance of the diode is reduced by using p- or n-type material of high conductivity, though the conductivity should not be so high that free-carrier absorption becomes at all noticeable. In a typical GaAs diode, the light is emitted through the window in a perforated metal electrode on the n-type side of the junction. The doping in the n-type region moves the Fermi level well into the conduction band so that the absorption edge is pushed further into the infrared region than for pure GaAs; the light emitted in the junction region is thus not reabsorbed. On the other hand, in the p-type region, the shift of the absorption edge (the so-called Burstein–Moss shift) is not nearly so pronounced, since the density-of-states effective mass of the holes is much greater than that of the electrons. The construction of a GaAs lamp is shown schematically in Figure 5.3a.

One of the most significant factors in reducing the efficiency of the GaAs lamp is total internal reflection of the radiation emitted from the junction. Like many semiconductors, GaAs has a rather high refractive index ($n = 3 \cdot 4$). Suppose we consider a large-area GaAs diode with a plane surface parallel to the junction. Then, if $\theta$ is the angle of incidence of light from the junction on to the surface, it will be totally reflected if $\theta > \sin^{-1}(1/n)$, where $n$ is the refractive index. Even when $\theta < \sin^{-1}(1/n)$, a considerable proportion of the light is reflected back into the diode, since the high refractive index implies a high reflection coefficient, even for normal incidence. To some extent the internal reflection loss can be reduced by making the front surface non-planar; the Weierstrass-sphere geometry is particularly favourable (Aigrain and Benoit à la Guillaume, 1957) in minimising total internal reflection. Nevertheless, bearing in mind such factors as absorption at the electrode on the front surface, one could hardly expect to reach an overall efficiency of 10% (Fischer, 1961).

Actually the quantum efficiency of a semiconductor lamp rises as the current input rises, since the non-radiative processes can become saturated when the densities of injected carriers are very large. High currents can only be passed through cooled crystals, so the efficiency is normally greater at liquid nitrogen temperature than at room temperature.

The advantages of the GaAs lamp include its mechanical stability, long life, and, by no means least, its capability of rapid modulation. Its ruggedness compared with the tungsten filament lamp has proved useful in tape reading and it has been employed in optical communication systems. The less efficient GaP lamp has been used in visual data displays and in film marking (typical photographic films have a spectral response that is not too different from that of the eye).

The internal efficiency of a GaAs electroluminescent junction is so high that it permits the lamp to be operated as a laser under certain circumstances (Hall *et al.*, 1962; Nathan *et al.*, 1962; Quist *et al.*, 1962). Firstly, it is necessary to provide a pair of plane parallel surfaces *perpendicular* to the junction. The geometry of the GaAs laser is shown in Figure 5.3b. Secondly, the injected current must become so large that population inversion occurs. When these conditions are satisfied the rather broad-band emission of the lamp becomes modified as shown schematically in Figure 5.4. The diode itself acts as a Fabry–Perot resonator (it is not necessary to coat the reflecting surfaces because of the high refractive index). The semiconducting laser emits a number of spectral lines each of which corresponds to an integral number of wavelengths between the mirror surfaces.

From what has been stated, it is clear that semiconducting lasers must be used with high currents, and hence, at high power. It is not surprising that GaAs lasers are usually operated under pulsed conditions or at liquid

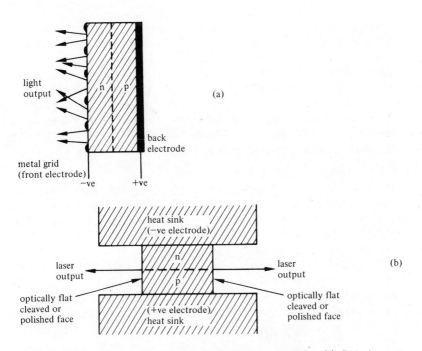

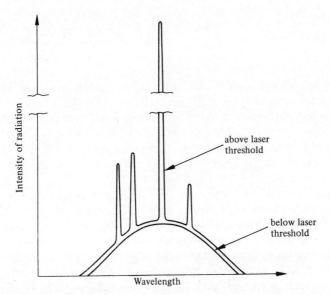

**Figure 5.3.** Junction electroluminescence in gallium arsenide: (a) GaAs lamp; (b) GaAs injection laser.

**Figure 5.4.** Spectral emission from a GaAs diode below and above the threshold for laser action.

nitrogen temperature. Nevertheless, it has been reported that a GaAs laser has been worked continuously at 200°K on a diamond heat sink (Dyment and D'Asaro, 1967). As compared with other lasers, the injection laser has the advantage of being very compact and it works from a simple power source. When considered merely as a light source, the advantage of the GaAs laser over the GaAs lamp is not only its greater monochromaticity or its greater power output. The laser output is normal to the surface of the crystal, so the efficiency is not limited by total internal reflection; overall efficiencies of the order of 50% become possible.

## 5.3 Electroluminescent layers and films

There are a number of applications where large-area electroluminescent panels are an advantage over small diode lamps or arrays of such lamps. Thus, since Destriau (1936) first demonstrated electroluminescence in phosphor layers, there has always been a great deal of interest in the development of electroluminescent panels.

The panels can be produced in two ways. Either particles of the phosphor (which is usually doped-ZnS) are embedded in a resin to form the so-called layer cell, or the phosphor is laid down as a thin film, usually by an evaporation technique. Since the films have not yet shown such a good performance as the layers, we shall discuss the latter in detail but, as will be seen later, thin-film electroluminescent panels offer certain advantages in principle, and may well supersede layer cells at some future date.

Layer cells have been made to operate from d.c. fields as well as a.c. fields, but it seems to be far easier to obtain a good performance using a.c. excitation, so we pay particular attention to this mode of operation. In particular, Fischer and his colleagues (1965) appear to have established the mechanism of a.c. electroluminescence in ZnS following a series of elegant experiments.

It must be noted that good photoluminescent or cathodoluminescent materials are not generally good for displaying electroluminescence, though ZnS can be used with all three types of excitation provided that it is suitably doped in each case. For electroluminescence, it seems to be essential to include a reasonable concentration of copper, probably more, in fact, than the amount that will enter into solid solution. Why copper is so important will become apparent from the model for electroluminesence that will be described.

In a typical layer cell, the active region consists of irregular copper-doped ZnS particles of perhaps 10 $\mu$m diameter contained in an insulating resin of 25-50 $\mu$m thickness as shown in Figure 5.5. The resin is chosen so as to have low losses and high dielectric strength. One of the electrodes is opaque but the other must be transparent; it can consist either of a very thin film of gold, or a rather thicker film of transparent semiconductor such as tin dioxide. Electroluminescence can be observed over a wide range of operating conditions; typically the cell may be subjected to a few hundred volts (a.c.) at a few hundred cycles per second.

It is easy to understand how visible luminescence can result from the simultaneous preśence of electrons in the conduction band and holes in the valence band. The holes will generally fall into deep traps before they can recombine with the electrons. Green or yellow luminescence in ZnS occurs when the electrons combine with the trapped holes. What is not so easily understood is the mechanism by which the electrons and holes are excited in the first place.

Electroluminescence can be seen with average electric fields of no more than about $10^6$ V m$^{-1}$, whereas it is considered that a field two orders of magnitude greater might be needed to excite free electron-hole pairs. In electroluminescent diodes the average field is certainly quite low, but there is intense field concentration at the depletion layer in the region of the p-n junction. One must look for some other field-concentrating mechanism in the layer cells.

One of the most striking features of Fischer's work has been his direct observation of the centres of electroluminescence in ZnS cells (Fischer, 1962). He employed so-called gap cells in which the active material is contained between two opaque electrodes and the luminescence viewed in a direction perpendicular to the electric field. Microscopic observations had been attempted before, but had proved almost worthless because of strong refraction at the surfaces of the irregularly shaped phosphor particles. ZnS has a very high refractive index ($n = 2 \cdot 37$) and one can only hope to look into particles of the compound without distortion if they are embedded in a medium of about the same refractive index. Fischer solved this problem by embedding his ZnS particles in an As-S-Br glass; the composition $As_{33}S_{30}Br_{37}$ has the same refractive index as ZnS. It was found that the electroluminescent centres were revealed as comet-like streaks of light, usually occurring in pairs with the narrow ends facing towards each other. Figure 5.5 shows the general effect (portrayed for

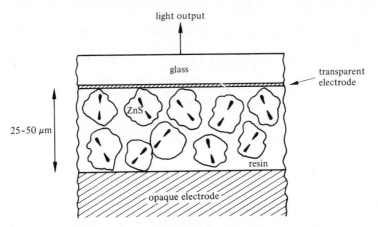

**Figure 5.5.** Section of an electroluminescent layer cell.

the layer-cell rather than the gap-cell geometry), although one can in practice see several pairs of 'comets' in many of the particles.

Fischer argued that the non-luminescent region between each pair of 'comets' is probably a linear inclusion of copper or $Cu_2S$ precipitated along a dislocation. Such inclusions are invisible provided that they are much narrower than the wavelength of light. $Cu_2S$ is a semiconductor with a much smaller energy gap than ZnS, while copper is a metal; thus we expect the field within the inclusions to be negligible compared with the field in the ZnS. The presence of fine well-conducting inclusions obviously gives rise to regions of exceptionally high electric field near their tips, and it is in these regions that the light appears.

Next, we must explain how the high electric field gives rise to luminescence. There are at least two satisfactory models that have been proposed. One model is based on the tunnel emission of electrons and holes from the ends of a semiconducting ($Cu_2S$) inclusion as shown in Figure 5.6. At one end holes are injected into the ZnS and soon become trapped, while the electrons are injected into the ZnS at the other end. When the polarity is reversed in the second half-cycle, electrons and holes are emitted respectively from the opposite ends of the inclusion; the electrons recombine with the previously trapped holes, emitting photons. As the figure shows, such a model requires that the electrons and holes tunnel across less than half the energy gap, the necessary field obviously being less than that for tunnel injection across the whole gap.

In the second model, illustrated in Figure 5.7, we can consider the inclusion to consist of metallic copper. When the field is in one direction, electrons are injected from the metal into the ZnS (across about one-half of the energy gap). Some of these electrons fall into shallow electron traps while others fall into holes that have been trapped during the

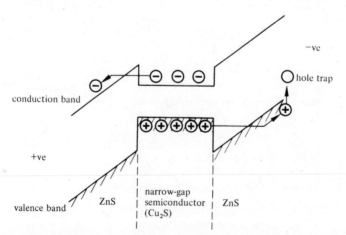

**Figure 5.6.** Model for excitation of a.c. electroluminescence involving tunnel injection of electrons and holes into ZnS.

previous half-cycle, giving rise to luminescence. At the other end of the inclusion, electrons that have been trapped during the previous half-cycle are released and accelerated by the high field causing impact generation of electron–hole pairs. The impact-generated holes are trapped almost immediately, while the electrons enter the inclusion replacing those that have been emitted from the other end.

Both these models explain the observed fact that at any time it is the 'comet' of each pair that is on the side nearest to the positive electrode that emits light. The 'comet' on the opposite side is luminescent during the opposite half-cycle. The models also explain why no light is emitted during the very first half-cycle when an alternating field is applied to a cell that has been rested; during this half-cycle there are no trapped holes with which the injected electrons can combine. Fischer and his coworkers (1965) have, in fact, demonstrated that the assumption of conducting inclusions and the adoption of one or other of the excitation models explain most of the experimental observations and are consistent with the remaining observations.

D.c. electroluminescence has been reported (Vecht *et al.*, 1968) to have been achieved using ZnS layer cells that are not too dissimilar from the a.c. cells that have been described above, though they generally require some sort of forming process (i.e. the passage of a moderately high current through the cell) before they can be used. It is, of course, possible for alternating current to flow through a layer cell in spite of the insulating resin, but direct current can pass only if there is some conducting path (or paths) between the two electrodes.

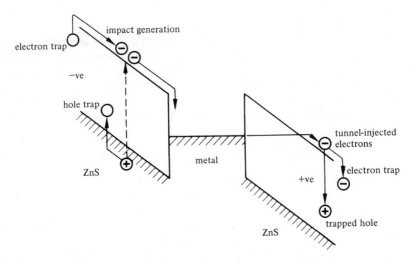

**Figure 5.7.** Model for excitation of a.c. electroluminescence based on tunnel injection of electrons and impact ionisation in a high field.

Neither of the excitation models that have been employed successfully for a.c. cells can be used to explain d.c. electroluminescence, since they do not provide a means for regeneration of carriers. It seems, however, that one might be able to explain d.c. electroluminescence assuming tunnel injection of electrons and holes respectively from neighbouring ends of different inclusions (rather than from opposite ends of a single inclusion). The electron emission and impact ionisation model could be adapted in the same way. As yet, however, there have been too few experiments on d.c. layer cells to permit much more than speculative theories to be formulated.

Layer cells of both the a.c. and d.c. types require the application of rather large voltages (greater than 100 V), whereas electroluminescent diodes can be operated from single-cell batteries. It would often be advantageous to have large-area panels (like the layer cells) that would work at low voltages. There seems to be no reason, in principle, why this should not be possible, but typical layer cells would seem to consist of a series-parallel arrangement of luminescent elements. What is needed, then, for low-voltage operation, is a means of cell fabrication that leads to a purely parallel arrangement. This is one of the principle reasons for studying cells in the form of thin films.

The insulating medium in which the particles of a layer cell are embedded can only add to the losses during operation. The fact that no such medium is needed in a thin-film cell provides an added incentive to research on electroluminescent films, particularly for d.c. operation (Thornton, 1962). Nevertheless, the behaviour of thin-film cells to date can only be described as disappointing.

### 5.4 Solid-state lasers

The semiconductor lasers that have already been discussed were preceded historically by optically pumped lasers, made from insulators doped with paramagnetic ions (Maiman, 1960). When one speaks of 'solid-state lasers', it is, in fact, these optically pumped devices that are usually implied. Solid-state crystalline lasers have a distinct advantage over injection lasers (and over gas lasers too) in that some of the metastable energy states have very long lifetimes (of the order of $10^{-3}$ s). The minority carrier lifetimes in the semiconductors that are used in injection lasers are more typically of the order of $10^{-9}$ s.

Population inversion may be achieved in the so-called three-level and four-level systems. In a three-level system, illustrated in Figure 5.8a, the active ions are excited from state 1 to state 3 by the pump (a bright source of incoherent light). If the states 2 and 3 are reasonably strongly coupled together, ions in state 3 relax towards state 2. Thus, when the pump power is sufficient, the population of ions in state 2 exceeds the number in state 1. This population inversion can produce laser action if the system is placed in a suitable optical cavity.

The three-level system suffers from the disadvantage that the stimulated transition involves the ground state (state 1), which is normally heavily populated. It is much easier to achieve population inversion between two states if the lower one is lightly populated, as in the four-level system shown in Figure 5.8b. In the example shown, the system is pumped from state 1 to state 4, and it relaxes from state 4 to the closely coupled state 3. Stimulated emission results from transitions between states 3 and 2, with spontaneous relaxation between states 2 and 1. In equilibrium, state 2 will be almost empty, provided that its energy is many times $kT$ above that of state 1. It may be necessary to cool the laser to achieve this condition.

Most of the earliest solid-state lasers were made from ruby, that is alumina $Al_2O_3$ doped with chromium at a level of a few hundred parts per million. The chromium ions ($Cr^{3+}$) occupy aluminium sites and are the active centres, their energy levels being determined by the crystal-field splitting of the host lattice. The energy states that are important in the three-level ruby laser are shown in Figure 5.9. The ions are pumped from the ground state $^4A_2$ to the $^4F_2$ band, which is at an energy of some 2 eV higher, using visible white light. The $^4F_2$ band is coupled to the $^2E$ level which has a lifetime of about $3 \times 10^{-3}$ s at room temperature. Thus it is not too difficult to obtain a high population in state $^2E$.[2] Stimulated emission between the $^2E$ state and the ground state has a wavelength of 0·6943 $\mu$m in the red region of the visible spectrum.

In the first solid-state laser, Maiman (1960) used a ruby of 1 cm diameter and 10 cm length. doped with 0·05% chromium. The end faces were plane polished and parallel to within 1 minute of arc. The optical cavity was formed by providing a completely reflecting surface at one end and a partially reflecting surface having 10% transmission at the other.

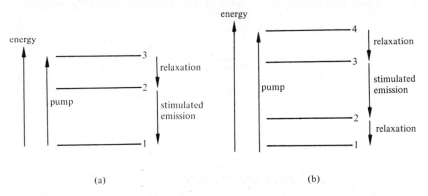

**Figure 5.8.** Energy diagram for (a) three-level laser and (b) four-level laser.

[2] Actually the 'level' $^2E$ really consists of a pair of levels separated by an energy difference of about $4 \times 10^{-3}$ eV, but this is ignored here.

This second surface allowed the feedback necessary for a growing in-phase signal as well as the emission of the laser light. In Maiman's laser, the pump was a coiled xenon flash tube, surrounding the ruby rod, giving an energy of 2000 J in pulses of a few milliseconds. More recently, the arrangement shown in Figure 5.10 has been adopted; here the flash tube and the laser crystal are placed at the foci of an elliptical reflector.

Aluminium oxide has many virtues as a host material; in particular, it has a high thermal conductivity and, although stable up to high temperatures, it can be grown in single-crystal form, either by flame fusion or, better, by pulling from the melt. The transition metals are easily incorporated substitutionally on the aluminium sites, but these sites

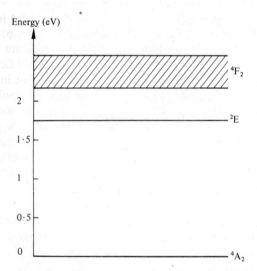

**Figure 5.9.** Energy states in ruby showing transitions involved in the laser.

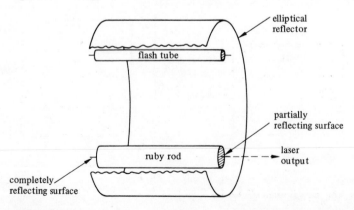

**Figure 5.10.** Arrangement of flash tube and laser crystal in a solid-state laser showing (cut away) elliptical reflector.

are not large enough to accommodate ions of the rare-earth metals. Since it is the rare-earth metals that provide the energy levels that are needed for operation in the four-level mode, and, since this mode can lead to laser action at lower pumping powers, there has been considerable interest in other host crystals. Good crystals of the fluorides of the alkaline earths can be grown by the Bridgman method and these crystals can take up trivalent rare-earth ions. Neodymium-doped $CaF_2$ is an example of such a crystal, giving a laser output at $1 \cdot 046$ $\mu m$ (Johnson, 1962). It has been found possible to operate a number of lasers, made from rare-earth-doped crystals, continuously rather than pulsed; for example, calcium tungstate $CaWO_4$ containing 1% $Nd^{3+}$ has given continuous laser output at room temperature at a wavelength of $1 \cdot 058$ $\mu m$ (Johnson et al., 1962).

Probably the best solid-state lasers at the present time are made from $Nd^{3+}$-doped yttrium aluminium garnet (YAG). It is most interesting to note that a laser made from this crystal has been used to produce visible light, this being achieved through the intermediary of a non-linear optical system (Giordmaine, 1969). The dielectric constant of a crystal changes if it is subject to an exceedingly strong electric field, such as the field in the electromagnetic radiation from a laser. Thus, if intense radiation at a frequency $\omega$ is incident upon such a crystal, the emergent radiation can be rich in components of frequency $2\omega$ and higher harmonics. The crystal $Ba_2NaNb_5O_{15}$ is strongly non-linear in its behaviour and has been incorporated in an Nd-YAG laser as shown in Figure 5.11. The output from an Nd-YAG laser usually has a wavelength of about $1 \cdot 06$ $\mu m$ but, in this case, both mirrors were almost perfect reflectors of such infrared light. The right-hand mirror was, however, transparent to visible light and allowed the harmonic at $0 \cdot 53$ $\mu m$, which was produced by the $Ba_2NaNb_5O_{15}$, to emerge from the system. The visible light output ($0 \cdot 5$ W) reported for such a combination was nearly 50% of the infrared output from the same laser crystal.

It is important that a laser material should be optically homogeneous: crystalline materials must, therefore, be grown with a high degree of perfection and with the impurities distributed perfectly uniformly. Optical homogeneity can be achieved rather more easily in glasses than in crystalline substances and numerous studies have been made of rare-earth-

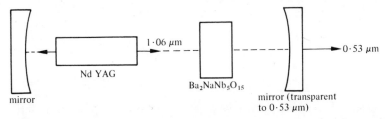

**Figure 5.11.** System for producing visible light from an infrared Nd-doped YAG laser using a harmonic generator.

doped glasses as laser materials (Snitzer, 1966). The advantages of glass as a host include its flexibility of size and shape, and its use allows one to select values for the temperature coefficient of the refractive index and strain-optic coefficients that make the number of wavelengths in the cavity temperature-independent. However, glasses have very low values for the thermal conductivity, so that glass lasers are not usually suitable for continuous operation or for pulsed operation at high repetition rates.

Very high pulses of power of short duration can be obtained from a laser by utilising the technique of $Q$-switching. In this technique, feedback is eliminated until the population inversion has reached an extremely high level. Then, when feedback is restored, a giant coherent pulse is emitted. $Q$-switching can be realised using the system shown in Figure 5.12. The optical cavity includes a polariser and a Kerr cell[3] as well as the laser rod. The Kerr cell and polariser normally prevent the return of light to

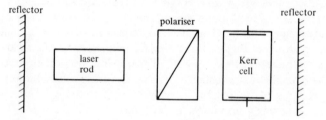

**Figure 5.12.** $Q$-switching a laser with an electro-optic element.

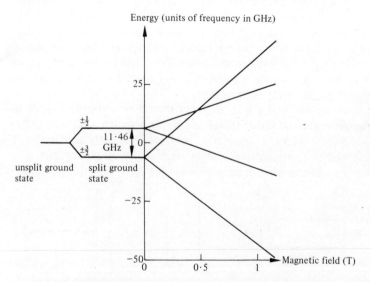

**Figure 5.13.** Splitting of the ground state of $Cr^{3+}$ in $Al_2O_3$ (magnetic field along trigonal axis).

[3] The principle of the Kerr cell is discussed later in Section 7.6.

the laser from the right-hand reflector, but this restriction can be removed when an electric field pulse is applied to the Kerr cell. An alternative method of $Q$-switching replaces the Kerr cell and polariser by a bleachable absorber. This substance interrupts the passage of light to and from the reflector until all its absorbent centres have been saturated; thereafter it is transparent.

Finally, a brief mention may be made of the solid-state maser, which actually appeared on the scene before the laser. Again ruby is a very widely used material but, of course, much more closely spaced energy levels are needed for operation at microwave frequencies. Figure 5.13 shows that the ground state of ruby really consists of two levels that are split, owing to the crystalline non-cubic field and spin–orbit coupling, by a small energy difference (corresponding to a frequency of $11 \cdot 46$ GHz). Each of these levels can itself be split by the application of a magnetic field, so that there are four spin states available. The ruby maser differs from the ruby laser in that it incorporates a magnetic field, it must be cooled to liquid helium temperature because of the small separation of the energy levels, and, of course, it makes use of a microwave cavity rather than a Fabry–Perot reflecting system.

### References
Aigrain, P., and Benoit à la Guillaume, C., 1957, *J. Phys. Radium,* **17**, 709.
Black, J. F., and Ku, S. M., 1966, *J. Electrochem. Soc.,* **113**, 249.
Brander, R. W., 1965, *GEC J. Sci. Tech.,* **32**, 15.
Destriau, G., 1936, *J. Chim. Phys.,* **33**, 620.
Dyment, J. C., and D'Asaro, L. A., 1967, *Appl. Phys. Letters,* **11**, 292.
Fischer, A. G., 1961, *Solid-State Electron.,* **2**, 232.
Fischer, A. G., 1962, *J. Electrochem. Soc.,* **109**, 1043.
Fischer, A. G., Shrader, R. E., and Larach, S., 1965, *Photoelectronic Materials and Devices,* Ed. S. Larach (Van Nostrand, Princeton), p.1.
Fok, M. V., 1962, *Opt. Spectr. (USSR) (Engl. Transl.),* **13**, 349.
Giordmaine, J. A., 1969, *Phys. Today,* **22**, 38.
Hall, R. N., Fenner, G. E., Kingsley, J. D., Soltys, T. J., and Carlson, R. O., 1962, *Phys. Rev. Letters,* **9**, 366.
Johnson, L. F., 1962, *J. Appl. Phys.,* **33**, 756.
Johnson, L. F., Boyd, G. D., Nassau, K., and Soden, R. R., 1962, *Proc. Inst. Radio Engrs.,* **50**. 213.
Kauer, E., 1965, *Philips Tech. Rev.,* **26**, 33.
Logan, R. A., Gershenzon, M., Trumbore, F. A., and White, H. G., 1965, *Appl. Phys. Letters,* **6**, 113.
Maiman, T. H., 1960, *Nature,* **187**, 493.
Nathan, M. I., Dumke, W. P., Burns, G., Hill, F. H., and Lasher, G., 1962, *Appl. Phys. Letters,* **1**, 62.
Pilkuhn, M. H., and Foster, L. M., 1966, *IBM J. Res. Develop.,* **10**, 122.
Quist, T. M., Rediker, R. H., Keyes, R. J., Krag, W. E., Lax, B., McWhorter, A. L., and Zeigler, H. J., 1962, *Appl. Phys. Letters,* **1**, 91.
Snitzer, E., 1966, *Appl. Opt.,* **5**, 1487.
Thornton, W. A., 1962, *J. Appl. Phys.,* **33**, 3045.
Thorp, J. S., 1967, *Masers and Lasers* (Macmillan, London).
Vecht, A., Werring, N. J., and Smith, P. J. F., 1968, *Brit. J. Appl. Phys.,* **1**, 134.

# Magnetic materials and devices

## 6.1 Introduction

A very important development in applied magnetism was the production of materials having high permeability and high magnetic moment, but with negligible electrical conductivity. Before that, magnetic materials were limited to the ferromagnetic elements and alloys. Although these were useful for permanent magnets, eddy-current losses due to the high electrical conductivity made them unsuitable for all but the lowest-frequency a.c. applications.

Eddy-current losses can be reduced by laminating the materials, as in transformer cores, but, since the laminations have to be made correspondingly thinner for higher-frequency usage, it is still impracticable to go much beyond the audio-frequency range in this way. During the 1930's the answer to the problem of providing high-frequency magnetic materials seemed to lie with the iron-dust cores, a natural extension of laminations. The iron is present in very finely divided form, held together by a suitable cement. The resultant magnetic moment is somewhat reduced, owing to the diluting effect of the cement, but the electrical resistivity is greatly increased. Operation up to a few megahertz is possible with these materials, but the frequency response could not be extended much further in this way.

The breakthrough came with the discovery of the useful properties of ferrites (Snoek, 1946) in the 1940's. Ferrites form a class of chemical compounds having the general formula $Fe_2O_3.MO$, where M is a divalent element such as zinc, nickel, magnesium, or even iron again. They are good electrical insulators, they have a reasonable magnetic moment and permeability, and they may be used under appropriate conditions at frequencies up to tens of gigahertz.

Ferrites can be used in straightforward applications requiring high permeability and low conductivity (radio inductors, for example) but one consequence of the low electrical conductivity leads to more interesting possibilities. Application of a magnetic field to any magnetic material causes a couple to be experienced by every molecular magnetic moment. Each moment is associated with angular momentum; so the immediate tendency is to precess about the applied magnetic field. Natural damping results in the precession gradually dying away as the magnetic moment 'spirals in' to its final position along the magnetic field (if we neglect thermal effects). If a suitable alternating magnetic field is also present, resonance can occur (Kittel, 1947), and the permeability of the material shows a typical resonance characteristic[1]. This is the basis of ferrite

[1] Resonance is not normally observed in bulk ferromagnetic materials because of the high electrical conductivity. This makes it difficult to get significant penetration by electromagnetic waves and also any tendency to resonate will be heavily damped.

devices used in microwave applications.

Ferrites find rather a different use in high-speed computer memories. The two stable states of ferrite-core magnetisation form the basis of a convenient binary memory element. Complete switching cycles, from one state to the other, and back again, can be achieved in under a microsecond.

However, even these switching speeds are becoming too slow for future requirements. Somewhat surprisingly, a binary memory element inherently capable of switching within just a few nanoseconds is formed simply from a small, very thin disc of unadulterated permalloy.

The thin magnetic film (Blois, 1955) demonstrates some of the fundamental properties of single-domain magnetic particles, and the chapter ends with a brief look at the application of these properties in the preparation of magnetic materials having either very large or small coercivities.

### 6.2 Microwave ferrites

A ferrimagnetic moment $\mathbf{m}$ in a magnetic field $\mathbf{H}$ experiences a torque $(\mu_0 \mathbf{m} \times \mathbf{H})$.[2] The resultant rate of change of $\mathbf{m}$ is given by $d\mathbf{m}/dt = \gamma\mu_0\mathbf{m} \times \mathbf{H}$, where $\gamma$ is the magnetogyric ratio, and the precession frequency $f_L$ (the Larmor frequency) equal to $\omega_L/2\pi$ is just $-\gamma\mu_0 H/2\pi$. By applying suitable alternating fields, equal in frequency to $f_L$, resonance can be excited, and heavy absorption of energy from the applied field occurs.

A typical sample of ferrite comprises a more or less random array of ferrimagnetic domains. There is a correspondingly random magnetic field distribution within the sample. Application of an external field will modify the internal field distribution but, unless the ferrite is uniformly magnetised (saturated) by the applied field, the resultant torques would cover a range of values, and a sharp resonance would not be observed.

In practice, the shape of the ferrite and the magnitude of d.c. bias field are chosen so that the ferrite is uniformly saturated. The initial equation can then be scaled up to $d\mathbf{M}/dt = \gamma\mu_0\mathbf{M} \times \mathbf{H}$, where $\mathbf{M}$ is the magnetic moment per unit volume. The Larmor frequency is approximately $1 \cdot 76 \times 10^{11}$ rad $T^{-1}$ $s^{-1}$.

The equation may be modified to include damping, which although small is not always negligible. Bloch and Bloembergen, Gilbert, and Landau and Lifshitz proposed slightly different ways of doing this. The distinctions between them refer mainly to individual advantages in particular situations, but for small damping they all give the same results.

[2] In other chapters, where we are primarily interested in magnetic fields in air or free space, we have elected to use the field $\mathbf{B}$ $(= \mu_0\mathbf{H})$, rather than the field $\mathbf{H}$, because of the convenience of its unit (T or Wb $m^{-2}$). Here, where we are more concerned with fields in magnetic materials, it is more generally convenient to work in terms of $\mathbf{H}$.

The Landau-Lifshitz equation is

$$\frac{d\mathbf{M}}{dt} = \gamma\mu_0\mathbf{M} \times \mathbf{H} - \frac{\lambda\mu_0\gamma}{M}[\mathbf{M} \times (\mathbf{M} \times \mathbf{H})] .$$

When the damping is small, $\lambda \ll 1$, the equation reduces to

$$\frac{d\mathbf{M}}{dt} = \gamma\mu_0\mathbf{M} \times \mathbf{H} - \frac{\lambda}{M}\left(\mathbf{M} \times \frac{d\mathbf{M}}{dt}\right),$$

which is Gilbert's equation. This equation may be used to find the resultant behaviour of $\mathbf{M}$ under any applied field conditions. For instance, suppose a small alternating field, $\mathbf{h} = \mathbf{h}_0\exp(jwt)$, is applied in the plane normal to a large d.c. bias field $\mathbf{H}_0$. Using Cartesian coordinates with $\mathbf{H}_0$ along the $z$ axis, $\mathbf{i}_z$ being the unit vector parallel to the $z$ axis, etc., the total applied field is

$$H_0\mathbf{i}_z + \mathbf{h} = H_0\mathbf{i}_z + h_x\mathbf{i}_x + h_y\mathbf{i}_y = \mathbf{H}$$

and the resultant moment is

$$M_0\mathbf{i}_z + \mathbf{m} = M_0\mathbf{i}_z + m_x\mathbf{i}_x + m_y\mathbf{i}_y = \mathbf{M} .$$

Since $H_0$ is a bias field sufficient to saturate the sample whereas $h$ is small ($H_0 \gg h$), it is reasonable to assume that $m \ll M$, and $M_0 \approx M$. Also, in the steady state, the system will show axial symmetry about the $z$ axis, so that $dM_0/dt$ will be zero. Further, as $\mathbf{h} = \mathbf{h}_0\exp(jwt)$, one expects solutions of the form $\mathbf{m} = \mathbf{m}_0\exp(jwt+\phi)$ etc., whence the operator $d/dt$ may be replaced where appropriate by $j\omega$.

Making all the above substitutions in Gilbert's equation leads to

$$j\omega(m_x\mathbf{i}_x + m_y\mathbf{i}_y) = \gamma\mu_0(M_0h_x\mathbf{i}_y - M_0h_y\mathbf{i}_x - m_xH_0\mathbf{i}_y + m_yH_0\mathbf{i}_x )$$

$$-\frac{\lambda}{M}\{[M_0\mathbf{i}_z + (m_x\mathbf{i}_x + m_y\mathbf{i}_y)] \times j\omega(m_x\mathbf{i}_x + m_y\mathbf{i}_y)\} .$$

The second term on the right-hand side may be simplified, by expanding and substituting $M \approx M_0$, to $-(j\omega\lambda)/(m_x\mathbf{i}_y - m_y\mathbf{i}_x )$. The equation may now be split into $\mathbf{i}_x$ and $\mathbf{i}_y$ components:

$$j\omega m_x = -(\omega_L + j\omega\lambda)m_y + \omega_m h_y$$

and

$$j\omega m_y = (\omega_L + j\omega\lambda)m_x - \omega_m h_x ,$$

where $\omega_m = -\mu_0\gamma M_0$, and, as before, $\omega_L = -\mu_0\gamma H_0$. Thus

$$m_x + jm_y = \chi_+(h_x + jh_y) \tag{6.1}$$

and

$$m_x - jm_y = \chi_-(h_x - jh_y) , \tag{6.2}$$

where $\chi_+ = \omega_m/(\omega_L + j\omega\lambda - \omega)$ and $\chi_- = \omega_m/(\omega_L + j\omega\lambda + \omega)$. Equations (6.1) and (6.2), in conjunction with $m_z = 0$, are the key equations to an understanding of the basic principles of microwave ferrite devices.

The equations may in fact be combined by solving for $m_x$ and $m_y$:

$$m_x = \chi h_x - j\kappa h_y,$$

$$m_y = j\kappa h_x + \chi h_y, \text{ where } \chi = \tfrac{1}{2}(\chi_+ + \chi_-) \text{ and } \kappa = \tfrac{1}{2}(\chi_- - \chi_+),$$

and, as above, $m_z = 0$. That is, in matrix form,

$$\begin{pmatrix} m_x \\ m_y \\ m_z \end{pmatrix} = \begin{pmatrix} \chi & -j\kappa & 0 \\ j\kappa & \chi & 0 \\ 0 & 0 & 0 \end{pmatrix} \begin{pmatrix} h_x \\ h_y \\ h_z \end{pmatrix}. \tag{6.3}$$

The relation between $m$ and $h$ is neatly summarised as $\mathbf{m} = \chi_p \mathbf{h}$, where $\chi_p$ is the Polder tensor (Polder, 1949). In the first-order treatment of microwave devices, the tensor equation is unnecessary, and the scalar equations (6.1) and (6.2) are usually adequate.

The crucial point about Equations (6.1) and (6.2) is that they indicate different scalar susceptibilities for circularly polarised waves of opposite rotation travelling along the direction of the d.c. bias field. The corresponding permeabilities are

$$\mu_- = \mu_0(1 + \chi_-) = \mu_0 \left( 1 + \frac{\omega_m}{\omega_L + j\omega\lambda + \omega} \right) = \mu'_- + j\mu''_-$$

and

$$\mu_+ = \mu_0(1 + \chi_+) = \mu_0 \left( 1 + \frac{\omega_m}{\omega_L + j\omega\lambda - \omega} \right) = \mu'_+ + j\mu''_+.$$

The real and imaginary parts of $\mu_-$ and $\mu_+$ are shown in Figure 6.1 for a constant value of $\omega$. The low-field parts of the curve are not given, since the ferrite will then cease to be saturated and the above analysis cannot apply.

There are three main points to note from the permeability curves. Circularly polarised waves of one sense of rotation will be almost completely unaffected by the ferrite, whereas waves having the other sense of rotation will either be strongly attenuated if resonance occurs, or, away from resonance, will be propagated, but at a different velocity, due to the different value of $\mu'$.[3]

The operation of microwave devices can be appreciated by considering any magnetic component of the wave in the plane normal to the d.c. bias

[3] The reader may be surprised by Figure 6.1 if anticipating values of $\mu'$ up in the hundreds or more. However, the permeability falls off at higher frequencies, and the values indicated in Figure 6.1 are quite typical.

field to comprise appropriate oppositely rotating, circularly polarised components (Soohoo, 1968).

For instance, a linearly polarised wave travelling along the d.c. bias field has a magnetic vector normal to the bias field, e.g. $h_x = h_0 \exp(j\omega t)$. This field may be considered to be the result of superimposing two circularly polarised waves $h_x + jh_y$ and $h_x - jh_y$. If the sample of ferrite is thick and resonance occurs, the positively polarised wave will be heavily attenuated, and only a negative circularly polarised wave will emerge. On the other hand, with a thin sample of ferrite, and away from resonance, attenuation will be small but the different velocities of the two waves will result in a Faraday rotation of the resultant linear polarisation.

Microwave isolators allow propagation in one direction, but heavily attenuate waves entering in the opposite direction. An isolator based on Faraday rotation is shown in Figure 6.2. The bias magnetic field, ferrite length, and geometry are chosen so that the overall linear rotation in travelling through the ferrite is just 45° with negligible attenuation. A wave entering from the left is rotated 45° and has the correct orientation to leave via the right-hand waveguide. A similar wave entering from the right is again rotated 45°, but in the same sense, and is therefore not suitably polarised for propagation through the left-hand guide, being 90° 'out'.

Thin resistive cards are used to prevent the non-transmitted wave from being reflected by the impasse. These are positioned as shown in Figure 6.2, and absorb energy from waves whose electric vector lies in the plane of the cards. In this way unwanted reflections are absorbed, but the wanted waves are unaffected (Langley-Morris, 1957).

Practical values of forward attenuation (insertion loss) might be around

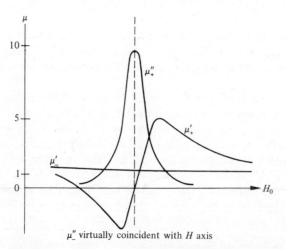

**Figure 6.1.** High-frequency variation of ferrite a.c. permeability with transverse bias field $H_0$.

1 or 2 dB, in devices designed for 30 GHz operation, with relative attenuation of at least 30 dB (that is, a power transmission ratio of 1000:1 between the two directions).

As a linearly polarised wave travels down a rectangular waveguide (Figure 6.3) there are two longitudinal planes $X$-$X$, $Y$-$Y$ in the waveguide where a stationary observer would experience an almost steadily rotating magnetic field. The direction of the rotation would depend on the direction in which the wave was travelling. An isolator based on resonance can thus be constructed, and is illustrated in Figure 6.4. A permanent magnet biases the ferrite transversely, and the field amplitude is set for resonance. Resonance, however, will only occur for waves travelling in the one direction, and such waves will, as a result, be attenuated, whereas waves travelling in the opposite direction will be unaffected (Sakiotis, 1956). The rotation is nowhere absolutely circular, but is optimum at about one-quarter the width of the waveguide from the edge. The advantage of this isolator over the Faraday rotation isolator is that it can be made physically shorter, since waveguide conversion sections (rectangular to circular and back again) are not needed.

The resonance isolator is, as might be expected, a narrow-bandwidth device. The field-displacement isolator is a broad-band device relying

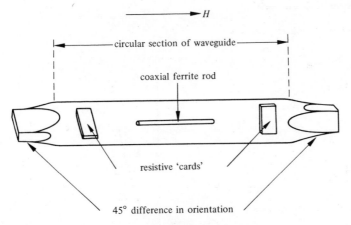

Figure 6.2. Faraday isolator for microwave use.

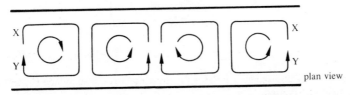

**Figure 6.3.** Plan view of magnetic field pattern of travelling wave in rectangular waveguide.

simply on the difference in permeabiliities experienced by waves travelling in opposite directions, considerably below resonance. If part of a waveguide is filled with ferrite, as in Figure 6.5a, any electromagnetic wave will redistribute itself dependent on the relative values of $\mu$ and $\epsilon$ of the ferrite and empty space. Since $\mu$ will differ for waves travelling in opposite directions, the electric field at the surface of the ferrite will also differ. Correct geometry, etc., can result in the two electric field intensities, for the two directions of propagation, being as shown in Figure 6.5b. A layer of resistive material on the appropriate ferrite surface will then absorb waves travelling in one direction, but will not affect those travelling in the opposite direction (Weisbaum and Seidel, 1956).

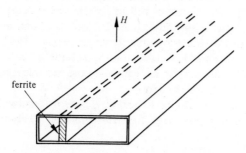

**Figure 6.4.** Resonance isolator.

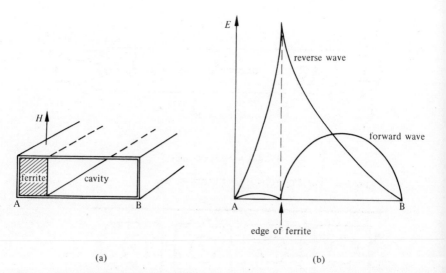

**Figure 6.5.** Field displacement isolator (a) showing ferrite loading and (b) showing electric field distribution.

Ferrite devices are of great value in radar systems. To ensure that the receiver aerial is aligned precisely with the transmitter aerial, it is usual to have only one aerial serving the two purposes. This must be connected to the transmitter in such a way that none of the transmitter power reaches the receiver directly and yet all of the received signal from the aerial must get to the receiver (see Figure 6.6).

A development of the Faraday rotation isolator achieves this. Side arms are attached as shown in Figure 6.7a, so that reverse waves are not absorbed but can leave by another path. A wave entering port 1 will pass straight through and out of port 2. A wave entering port 2 will, as above, be incorrectly polarised for leaving via port 1 but can excite an outgoing wave leaving via port 3. Similarly a wave entering at 3 leaves via port 4, and one entering at 4 leaves via port 1. The usual schematic representation of this microwave component is shown in Figure 6.7b. Not surprisingly it is called a circulator (Hogan, 1956), and is a particular example of a gyrator. Its use in radar systems would be as shown in Figure 6.8.

Although ferrite devices are particularly suited to present-day microwave frequencies, attempts are being made to raise the upper frequency limit and drop the lower limit. Higher frequencies mean greater magnetic fields and external magnets become impracticably large. The most promising approach seems to lie in selecting materials with high crystal anisotropy; ferrites are available with anisotropy fields of 2 T, and it may prove possible to use antiferromagnetic materials with internal fields of 10 T.

The difficulty about operation at lower frequencies (hundreds of megahertz) is that lower bias fields mean incompletely saturated ferrites, and hence the presence of domain walls. These can move in sympathy with the electromagnetic wave, and cause attenuation of all waves through wall-hysteresis effects.

One approach is simply to use large bias fields to ensure that the ferrite is saturated, and to accept that operation is then well below the resonant frequency. This limits the possible types of device, and the overall efficiencies obtained are not too good, since the difference between $\mu'_+$ and $\mu'_-$ is comparatively small.

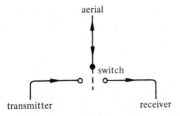

**Figure 6.6.** Basic transmit/receive switching requirement in a radar system.

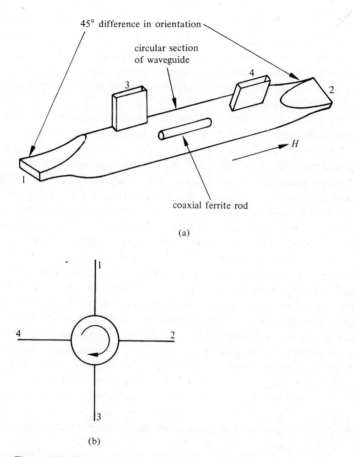

45° difference in orientation

circular section
of waveguide

coaxial ferrite rod

(a)

(b)

**Figure 6.7.** Faraday gyrator: (a) basic device; (b) microwave schematic representation.

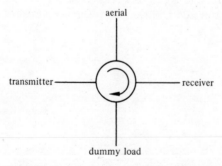

aerial

transmitter

receiver

dummy load

**Figure 6.8.** Use of a gyrator in a radar system.

## 6.3 Ferrite memory elements

The basic attraction of electronic computers is their very high speed. Because of the speed, a great deal of computation can be carried out in a few seconds, and larger jobs, that might occupy months or years by manual computation can be completed in hours or days. Not surprisingly the quest is always for yet faster computers, most of the effort being aimed at increasing the speed of the slowest part of the system, namely the computer memory.

During the operation of the computer, the memory will be consulted (read), or written into, during almost every operation within the computer, and a long memory cycle time means that the remainder of the computer has to wait for the memory either to absorb or to give out the desired information.

Modern electronic computers need bistable elements in great numbers to make up the computer memory. The most common memory element at present is the ferrite core. This is a small toroid of ferrite, usually some hundredths of an inch in diameter, which has two stable magnetic states. One state may represent a stored 'one' and the other, a stored 'zero' (see Figure 6.9). The high stability of each state is due to the low magnetisation energy, since ideally there will be no 'free poles'.

The cycle time of ferrite core memories is limited basically by the time taken to produce necessary changes in the direction of ferrite magnetisation. The limitation involves the characteristics of individual cores, but also arises partly in an indirect way, because a core memory comprises a very large number of cores rather than just one or two.

Magnetisation reversal within a ferrite core takes place by the movement of domain walls. If a current is passed along a wire which goes through the toroid, a circumferential magnetic field will be produced. Dependent on the direction of the current, the field will either confirm the direction of ferrite magnetisation, or will try to reverse its direction. Suppose the field is opposing the magnetisation. As the ferrite cores are made from square-loop materials [that is materials with square $(B, H)$ characteristics], nothing will happen unless the applied field $H_{app}$ is at least equal to the coercivity $H_c$ of the ferrite. When $H_{app}$ is equal to $H_c$, small reverse domains can start to increase in size by domain wall movement. As $H_{app}$ is increased above $H_c$, the domain walls will move right through the ferrite, and the whole magnetisation is 'switched' into the new direction.

'one'

'zero'

**Figure 6.9.** Diagram of a ferrite core showing the two stable magnetic states.

Operation of a single-bit ferrite core memory can be achieved with two conductors threaded through the core as in Figure 6.10. A comparatively large drive current (about 1 A) passed along the drive conductor in the appropriate direction writes in a 'one' or a 'zero' as desired. Read-out is obtained by passing a single current pulse down the drive conductor, but always in the same direction. Dependent on whether a 'one' or a 'zero' is stored, there will, or will not be, a change in ferrite magnetisation. If there is a reversal, there will be a corresponding flux change, which, being linked to the sense conductor, will produce a 'read' signal. If there were no magnetisation reversal, there would be no 'read' signal. The electronic circuitry associated with the store distinguishes between a 'read' signal and no 'read' signal, and can thus decide whether a 'one' or 'zero' has been stored. Clearly, at this stage the original stored information has been lost, since the cores are all bound to be in the same final state. Provision must be made for rewriting the store with either the original information, or with new information, as necessary.

Evidently the domain wall velocity is important if fast operation is desired. Experiments show that, to a reasonable degree of accuracy, the velocity is proportional to $H_{app} - H_c$, and the switching time $\tau$ is then inversely proportional to $H_{app} - H_c$. By making $H_{app}$ sufficiently large it would appear that $\tau$ could be made as small as one might wish.

The restriction arises because computer memories normally need sufficient capacity to store thousands of complete binary numbers (words) each involving, perhaps, sixteen binary bits. Each individual core then has to be integrated into the memory with some means of selecting (addressing) particular cores as the need arises.

Selection in a computer memory could be effected, in principle, by having an individual set of conductors to each ferrite core. Every core could then be operated independently of the others and high speed operation might be possible. However, even in a small store of, for example, a thousand words each of sixteen bits, this would involve some 16 000 sets of conductors, each having their own individual operating circuits. This is not a practical proposition!

Considerable economy in wiring and circuitry results if the cores are wired in rectangular arrays or matrices, as in Figure 6.11. Selection of a

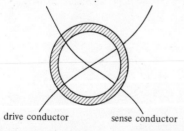

drive conductor          sense conductor

**Figure 6.10.** Wiring of a simple one-bit memory core.

single particular core is then possible if currents of suitable amplitude are passed down the appropriate $x$ and $y$ wires (Forrester, 1951). Equal currents are normally used in each wire, and to avoid disturbing unaddressed cores, it is clear that the field due to the current through a single wire must be somewhat less than $H_c$. The total field at the selected core is therefore limited to less than $2H_c$. Typically the minimum switching time is some hundreds of nanoseconds.

For many purposes this has been fast enough; the difficulties in designing the associated read and write circuitry have resulted in overall logic speeds compatible with the core switching speeds, giving complete computer cycle times of the order of a microsecond.

However, there is always a demand for faster computers and recent advances in electronics have resulted in considerably faster circuit operation. Unfortunately there has been no corresponding increase in domain wall mobilities.

The use of smaller ferrite cores is a reasonable way of attempting to reduce switching times, since domain walls will then not have so far to travel (Werner *et al.*, 1967). However, smaller ferrite cores introduce fresh complications, mainly owing to the use of finer wires to thread through them.

Actual wiring is, of course, made more difficult, but more serious is the increase in heat generation. Current amplitudes will be the same as for larger cores, but the currents are now passing through wires of smaller cross section. The rise in temperature may adversely affect the ferrite properties, or, in the limit, can lead to the wires fusing.

Another obstacle to progress comes from the relationship $V = L \, dI/dt$ where $L$ is the inductance of the core wiring. Faster stores mean larger values for $dI/dt$ and higher values of voltage. As the frequency response of the drive transistors is pushed up ($f_T$ going towards 1 GHz), so also must the voltage ratings be raised. Coupled with the need for currents of the order of 1 A, it can be seen that availability of suitable drive transistors for core stores may prove elusive, or expensive.

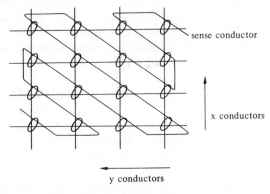

sense conductor

x conductors

y conductors

Figure 6.11. Ferrite-core memory matrix.

Smaller ferrite cores, and the more subtle use of cores of a given size, can reduce the cycle time to under half a microsecond, but a formidable price has to be paid in increased complexity (Russell *et al.*, 1968).

## 6.4 Thin magnetic films

Thin magnetic films are of somewhat specialised interest in that they form the basis for one type of ultra-fast computer memory element. However, a description is of wider significance because the thin magnetic film demonstrates, on a macroscopic scale, some of the characteristics typical of the microscopic crystallites that make up bulk ferromagnetic materials.

The crucial feature of thin magnetic films is that their thickness is sufficiently small for them to exist as single magnetic domains. This is not usually possible in bulk material because of the high magnetostatic energy that would be involved. As a result, large samples break up into almost randomly oriented domains, with minimal magnetostatic energy. However, the domain walls themselves have an energy content, proportional to the total area of wall. If the thickness of a large sample of material is steadily reduced, the magnetostatic energy, proportional to field strength squared, falls as the square of the thickness, whereas the total domain wall energy falls linearly with thickness. There is a limiting thickness at which the two balance and, below this thickness, the film of material can exist as a single magnetic domain.

Apart from small variations in composition, or the addition of one or two percent of other elements, permalloy is used in all thin film stores. The critical thickness for single domain stability is about $3 \times 10^{-7}$ m, and films are typically made around $10^{-7}$ m thick. They are normally circular or oval, about $10^{-3}$ m across, and deposited on an aluminium substrate in a rectangular array. The films are electrically insulated from the substrate by a thin layer of silica.

In practical applications one might simply have chosen material with the highest value of saturation magnetisation, but this has to be coupled with a low magnetostriction characteristic. Since the films must be deposited on a supporting substrate, any changes in ambient temperature introduce stresses due to inevitable differences between the expansion of the substrate and the film. It is very convenient that ordinary permalloy (an 80:20 nickel:iron alloy) is suitable, having a zero magnetostriction coefficient.

To drive the films, magnetic fields may be applied in the plane of the films, by passing currents along strip conductors laid on top of the rows and columns of films. The return currents will appear to pass along reflections of the drive conductors in the conducting substrate.

As described above, the film would have no inherent bistability, and this must be provided if the film is to be used as a binary memory element. Preparation is usually by vacuum deposition or chemical deposition and, by having the deposition take place in an externally applied magnetic field parallel to the plane of the films, a built-in uniaxial anisotropy results. In use, the film magnetisation, if left to itself, will lie along this easy direction equally readily one way or the other.

The uniaxial anisotropy energy $U$ is proportional to $\sin^2\theta$, i.e. $U = k\sin^2\theta$ (see Figure 6.12a), where $k$ is an important constant of the film material. Several methods are used in practice to determine the constant $k$. For instance, the film may be placed in a torque magnetometer, a very large static field **H** applied in the plane of the film, and the necessary torque to keep the easy axis of the film at known angles $\alpha$ to **H** measured. Then the torque is given by

$$\tau = MH\sin(\alpha-\theta) = \frac{\mathrm{d}U}{\mathrm{d}\theta} = k\sin 2\theta$$

at equilibrium. If $H$ is very large, $\sin(\alpha-\theta)$ must tend to zero, i.e. $\theta \to \alpha$, or, as expected, **H** pulls **M** into line with itself. In this case,

$$\tau = k\sin 2\alpha ,$$

whence $k$ may be found. Similarly, by varying **H**, **M** may be determined.

In computer technology, the ease with which **M** can be rotated normal to the easy direction is very important. If a d.c. field $H_\perp$ is applied, as in Figure 6.12b, the energy of the system (relative to that at $\theta = 0$) with **M** at an angle $\theta$ to the easy direction is given by

$$U = U_{\theta = 0} + k\sin^2\theta - MH_\perp\sin\theta .$$

In equilibrium, $\mathrm{d}U/\mathrm{d}\theta = 0$, whence,

$$2k\sin\theta\cos\theta - MH_\perp\cos\theta = 0$$

and

$$\sin\theta = \frac{M}{2k} H_\perp .$$

In other words, as $H_\perp$ increases from zero, the projection of **M** parallel to $H_\perp$ ($= M\sin\theta$) increases linearly with $H_\perp$ until, when $H_\perp = 2k/M$, **M** lies at

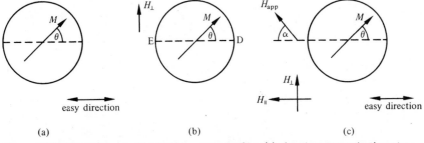

**Figure 6.12.** Plan view of uniaxial thin magnetic film (a) showing magnetisation at an arbitrary orientation to the easy direction, (b) showing a magnetic field $H_\perp$ normal to the easy direction, in the plane of the film, and (c) showing an arbitrary field $H_{\mathrm{app}}$.

right angles to the easy axis. The field strength $2k/M$ occurs frequently as a parameter in thin film work and is often termed the anisotropy field, $H_k$.

Suppose an arbitrary field $H_{app}$ is applied as in Figure 6.12c. The energy of magnetisation is now

$$U = U_{\theta=0} + k\sin^2\theta + MH_{\parallel}\cos\theta - MH_{\perp}\sin\theta \ , \tag{6.4}$$

where $H_{\perp}$ and $H_{\parallel}$ are respectively the components of $H_{app}$ normal and parallel to the easy direction. The equilibrium values of $\theta$ are given by equating $dU/d\theta$ to zero. A further differentiation determines whether the equilibrium positions are stable or unstable:

$$\frac{d^2U}{d\theta^2} = 2k\cos 2\theta - MH_{\parallel}\cos\theta + MH_{\perp}\sin\theta \ . \tag{6.5}$$

Clearly, when $H_{app} = 0$ ($H_{\perp}$ and $H_{\parallel}$ both zero) there will be two stable states, $\theta = 0$ and $\theta = \pi$. As $H_{app}$ increases, $d^2U/d\theta^2$ will decrease for one of these, and increase for the other. At a critical value of $H_{app}$, the former value of $d^2U/d\theta^2$ will become negative indicating that that originally stable orientation has become unstable. The magnetisation will then 'switch' to the other orientation. This will be an irreversible switch over, and subsequent reduction of $H_{app}$ to zero will leave **M** oriented along the easy direction, but in the opposite sense to that at the start. Figure 6.13 shows the variation of $U$ as a function of $\theta$ for various amplitudes of $H_{app}$, but at constant $\alpha$.

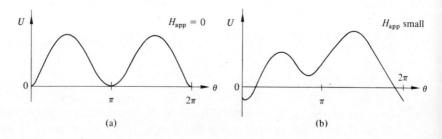

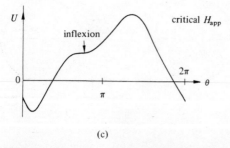

**Figure 6.13.** Variation of magnetic potential energy $U$ with magnetisation orientation: (a) zero applied field; (b) small applied field; (c) critical applied field strength.

The critical amplitude of $H_{app}$ at which switching can occur is a function of $\alpha$. A plot of critical field strength against $\alpha$ is shown in Figure 6.14. Its equation may be found by equating both $dU/d\theta$ and $d^2U/d\theta^2$ to zero, the result being

$$\left(\frac{H_\perp}{H_k}\right)^{2/3} + \left(\frac{H_\parallel}{H_k}\right)^{2/3} = 1 \ .$$

The symmetrical curve is called an astroid, and is of great value in thin film work (Smith, 1958).

If any field represented by the vector OA (Figure 6.15) is applied in the plane of the film, it can be shown that equilibrium positions of M are always tangential to the astroid from A, that is either PA or QA dependent on the initial direction of M along the easy direction. As OA increases, so the two possible orientations of M alter, until the applied field strength reaches the astroid. At OB there are still two equilibrium positions BR and SB, but on any further increase in applied field, to OC, for example, there is only one, namely TC. A film with initial magnetisation along OX will switch irreversibly with this increase in field, from BR to TC. Subsequent reduction of the applied field results in M falling back to the easy direction (TC to SB to QA to OY).

Experimental measurements of the speed at which the film magnetisation can be switched yield results of the order of only 2 to 3 ns (Dietrich and Proebster, 1960). Two factors contribute to this very high speed. The magnetisation of the whole film rotates together, coherently. This is in contrast with ferrite core switching where domain walls have to travel right through the material. Since movement of a domain wall implies magnetisation rotation within the wall, it is inevitable that switching by domain wall motion must be slower than by coherent rotation. The

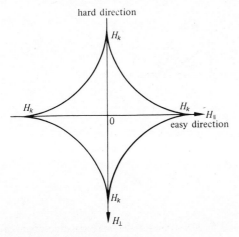

**Figure 6.14.** The astroid.

second factor arises from the shape anisotropy of the film. A field in the plane of the film, and normal to the easy direction, produces a torque on **M** which in the first instance tends to lift **M** out of the plane of the film (**T** = **M** × **H**). The large demagnetising field normal to the plane then results in a second torque, much larger than the first, that rotates **M** very rapidly within the plane of the film. In effect, **M** is prevented from precessing by shape anisotropy, and so rotates to its final equilibrium position via a flat elliptical path (Figure 6.16). Under these conditions oscillatory overshoot would be expected, and has been observed in practice (Werner *et al.*, 1967). (A similar effect is observed if a toy mechanical gyroscope is prevented from precessing about a vertical axis by a firmly held obstacle. As soon as the precessing gyroscope reaches the obstacle it experiences a 'shape anisotropy' torque preventing its precession, and the gyroscope rotates rapidly downwards to the vertical.)

The ideal switching threshold is given by the astroid, Figure 6.14, but attempts to confirm the astroid using quasi-static magnetic fields result in a threshold similar to that given by Figure 6.17. Closer investigation shows that this is because switching can occur by domain wall motion at field strengths less than those necessary for coherent-rotation switching, given sufficient time. Even though such domain wall switching would be much slower, it is nevertheless cumulative, and because, in practice, $H_c < H_k/2$ (typically $\mu_0 H_k \sim 4$ or $5 \times 10^{-4}$ T, $\mu_0 H_c \sim 1$ or $2 \times 10^{-4}$ T), it is impossible to operate a thin film store on a coincident-current selection basis as with ferrite cores.

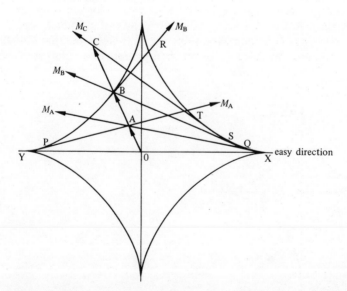

**Figure 6.15.** The use of the astroid in determining stable magnetisation orientation under arbitrary field conditions.

Thin magnetic film computer memories usually work in a word-address mode (Raffel, 1961) as shown in Figure 6.18. The sequence of operation is shown in Figure 6.19. It begins with a word drive current (Figure 6.20) producing a field greater than $H_k$ at all the films along the drive conductor. The magnetisation vectors rotate to the hard direction, either clockwise or anticlockwise dependent on the initial orientation.

The rotation of magnetisation results in a change in flux associated with all the 'sense' conductors, and a brief signal is induced in them. The

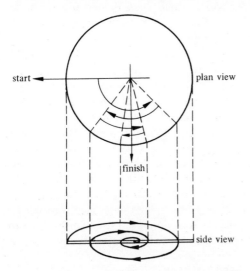

**Figure 6.16.** Variation in thin-film magnetisation orientation following the sudden application of a hard-direction field $H_\perp$ greater than $H_k$.

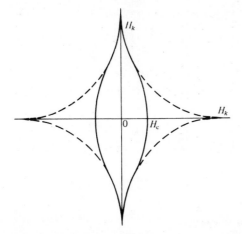

**Figure 6.17.** Practical thin-film switching threshold.

polarity of this signal depends on whether the magnetisation rotation was clockwise or anticlockwise, and reflects the original information stored in the element.

As with coincident-current operation, read-out is a destructive process, and it is necessary to rewrite the store. Before the word drive current is reduced, comparatively small digit currents ($H < H_c$) are passed along the digit conductors. The direction of these currents determines the information to be rewritten into the store by tilting the magnetisation slightly from the hard direction, so that when the word drive is removed the magnetisation will relax to the appropriate orientation along the easy direction. Finally the digit currents are removed.

In every such cycle all the films under a particular drive conductor are read out and rewritten. If the computer is designed to operate with a twenty-bit word length, every drive conductor will be just twenty films long. Addressing (selecting) a particular drive conductor then, results in finding a complete word (hence the name, word-address mode).

In principle, the operation and construction of thin film memories are straightforward. If the films are made by deposition from vacuum, it is practicable to prepare entire matrices of, for example 64 × 48 films, in one operation. Typically an aluminium substrate, 50 × 75 × 5 mm³ in size, is first coated with an electrically insulating layer of silicon monoxide, $3 \times 10^{-5}$ m thick. The array of films is then deposited through a suitably perforated mask. Printed-circuit techniques are used to produce the parallel conductor patterns on thin insulating material, and these are gently

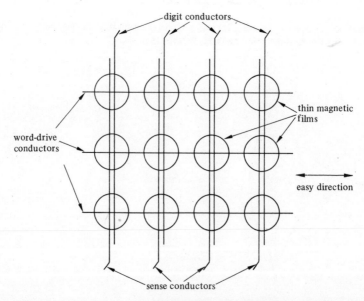

**Figure 6.18.** Wiring of a word-address thin-film matrix.

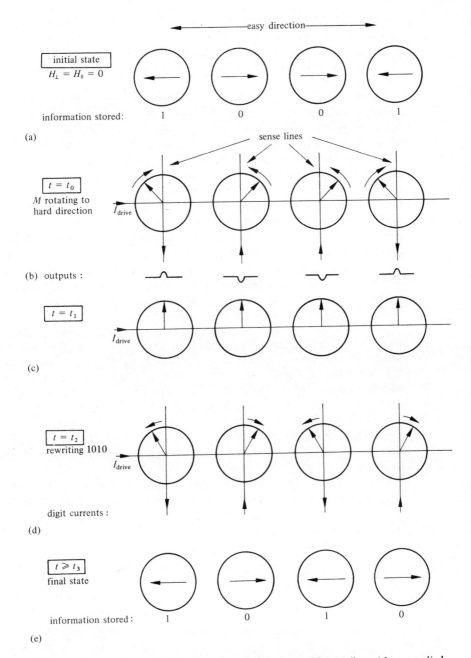

**Figure 6.19.** Operation of a four-bit word address store: (a) initially, with no applied fields; (b) during the application of a hard-direction field $H_\perp$; (c) system steady with $H_\perp > H_k$; (d) digit field applied in easy direction, $H_\perp$ still on; (e) final state, after removal first of $H_\perp$, then of the digit field.

pressed on top of the film matrix. Because the conductors are in the form of transmission lines, they appear to the associated circuitry simply as resistive loads, and inductive effects are completely eliminated. Also, since the strip conductors are of reasonable cross-section area, there are no heating problems.

Present thin-film memories designed for high-speed operation have cycle times around the 100 nS mark (Anacker *et al.*, 1966), but some thought has been given to larger stores with a less spectacular speed where cheapness of the thin film would be the main attraction (Raffel, 1964).

The basic ideas associated with thin magnetic film memory elements are elegant, but in practice there are difficulties. For instance, the transfer efficiency of the film, looked on as a means of coupling the drive power to a sense system, is extremely low. Volts of drive are required, and only a millivolt or so of signal appears. Nevertheless, amplifying techniques are now such that this is not an insuperable difficulty. The main stumbling block at present seems to be the low yield of completely satisfactory thin film matrices. Clearly, even if only one film on a matrix of 64 × 48 is unacceptable, then the whole matrix is useless. In contrast with what can be done with ferrite core matrices, there is no chance of picking out the faulty element and replacing it with a good one.

An alternative approach to thin films is the cylindrical film, or plated wire. The film is chemically deposited as a coating on a single combined sense/digit conductor, and the drive conductor is then placed around the films as shown in Figure 6.21. The cylindrical film is inherently bistable, with the magnetisation circumferentially round the cylinder one way or the other (as the ferrite core). Operation is basically as described above, for the flat film; the word drive rotates the magnetisation parallel to the cylindrical axis, inducing an output signal, and a digit current can then write desired information in by slightly tilting the magnetisation towards the 'easy direction'.

The advantage of the cylindrical film is that there is in principle no magnetostatic energy associated with the closed circumferential flux path,

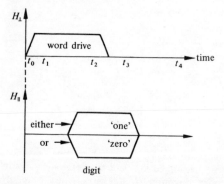

**Figure 6.20.** Field sequence during word-address operation.

and hence the film thickness may be increased well above that of the flat film, resulting in much larger output signals. On the other hand, the cylindrical film is not as amenable as the flat film to transmission line conductors and cycle times are more modest (Fedde, 1967).

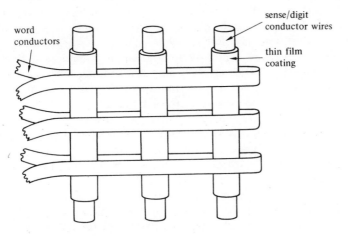

**Figure 6.21.** Cylindrical film memory.

## 6.5 Magnetic alloys

The saturation moment of a magnetic alloy can usually be predicted from a knowledge of its composition and crystal structure. However, by suitable metallurgical treatment, a particular alloy can be given either a high or low coercivity, and, by implication, higher or lower values of $(BH_{max})$. Thus, considerable research is aimed at increasing the range of magnetic coercivities available with particular materials.

The thin magnetic film showed that reversal of magnetisation by movement of domain walls can occur at lower field strengths than those needed for coherent rotation. For low coercivity in bulk material it is necessary to encourage the existence of domain walls and to facilitate their movement.

Domain wall motion is hindered by irregularities in the crystal structure of the material. If the wall is to pass by an irregularity, there is an intermediate stage when the wall in some way or another must straddle the imperfection. [Figure 6.22 is a considerable simplification of what happens (Middlehoek, 1961).] A certain amount of wall energy has to be replaced by a different amount of energy, and, depending on the relative magnitudes of these energies, the domain wall either will be reluctant to straddle the imperfection, or will be reluctant to leave it. In both cases movement of the domain wall has been hindered and the coercivity raised.

For low coercivity, then, metallurgical treatment is aimed at producing large crystallite sizes, too large to exist as single domains, and at producing as perfect a lattice as possible within each crystallite, so that the domain walls can move comparatively freely.

Quite the opposite features must be encouraged in high-coercivity materials. Domain walls, or at least their movement, must be suppressed, and reversal of magnetisation by rotation must then be made as difficult as possible. A comparatively simple way of achieving this is by deliberately introducing irregularities, either by appropriate metallurgical methods, or by adding impurities. Domain walls may still abound, but their movement is made virtually non-existent. The resultant coercivity is then directly related to the inherent crystal anisotropy of the alloy.

Reliance on crystal anisotropy for high coercivity has a possible disadvantage; rotation coercivity is inversely proportional to saturation magnetisation (as, for example, in the thin magnetic film where $H_k = 2k/M$). Crystal anisotropy may be of more significance where a high $H_c$ is needed and a comparatively low $M$ can be tolerated.

A more effective approach is to prepare the magnetic material itself in the form of microcrystallites, each of which is sufficiently small to exclude the existence of domain walls, and then to control the anisotropy of the individual crystallites. Crystallite anisotropy is determined by two factors, the inherent crystal anisotropy, as above, and the anisotropy associated with the individual shape of each crystallite. Shape anisotropy increases with $M$, and high coercivity will result if the individual crystallites are in the form of long thin filaments, reasonably parallel to a specified direction. If suitable alloys are chosen, solidification from the melt can result in two phases, one of which forms the desired filaments, correctly oriented. The expense of such magnetic materials reflects the difficulty of finding suitable alloys, and the complexity of the metallurgical processes involved.

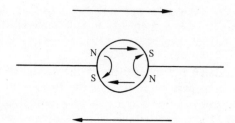

**Figure 6.22.** Simplified view of domain wall passing an irregularity.

## References

Anacker, W., Bland, G. F., Pleshko, P., and Stuckert, P. E., 1966, *IBM J. Res. Develop.*, **10**, 41.

Blois, M. S., 1955, *J. Appl. Phys.*, **26**, 975.

Dietrich, W., and Proebster, W. E., 1960, *J. Appl. Phys.*, **31**, 281S.

Fedde, G. A., 1967, *Electronics*, **40**, number 10, 15/5, 101.

Forrester, J. W., 1951, *J. Appl. Phys.*, **22**, 44.

Hogan, C. L., 1956, *Proc. Inst. Radio Engrs.*, **44**, 1345.

Kittel, C., 1947, *Phys. Rev.*, **71**, 270.

Langley-Morris, A., 1957, *Proc. Inst. Elec. Engrs. (London), Pt.B, Suppl.6*, **104**, 383.

Middlehoek, S., 1961, *Ferromagnetic Domains in Thin NiFe Films* (North Holland Publishing, Amsterdam), p.40.

Polder, D., 1949, *Phil. Mag.*, **40**, 99.

Raffel, J. I., 1961, *Proc. Inst. Radio Engrs.*, **49**, 155.

Raffel, J. I., 1964, *J. Appl. Phys.*, **35**, 748.

Russell, L. A., Whalen, R. M., and Leilich, H. O., 1968, *Inst. Elec. Electron. Engrs., Trans. Magnetics*, **MAG-4**, 134.

Sakiotis, N., 1956, *Inst. Radio Engrs., Trans. Microwave Theory Tech.*, **4**, 240.

Smith, D. O., 1958, *J. Appl. Phys.*, **29**, 264.

Snoek, J. L., 1946, *Philips Tech. Rev.*, **8**, 353.

Soohoo, R. F., 1968, *Inst. Elec. Electron. Engrs., Trans. Magnetics*, **MAG-4**, 118.

Weisbaum, S., and Seidel, H., 1956, *Bell System Tech. J.*, **35**, 877.

Werner, G. E., Whalen, R. M., Lockart, N. F., and Flaker, R. C., 1967, *IBM J. Res. Develop.*, **11**, 153.

# Dielectricity

## 7.1 Dielectrics and ferroelectrics

Any material placed in an electric field becomes polarised, that is it acquires an electric dipole moment and the field strength within differs from that outside. In situations where the polarisation and field changes are practically important, the material is usually referred to as a dielectric.

The difference in field strength may be found by using the Gauss law, which states that, for any volume $dv$ with an enclosing surface $dS$,

$$\int_{\text{surface}} \mathcal{E} \, dS = \int_{\text{volume}} \left( \frac{\rho - P}{\epsilon_0} \right) dv \, ,$$

where $P$ is the volume polarisation and $\rho$ the net charge density. In dealing with dielectrics, however, this law is not in its most useful form since $P$ is a function of $\mathcal{E}$.

Gauss's law is, of course, still true if the volume embraces a surface boundary, and leads to the well-known pill-box technique for relating internal and external field strengths. For example, in terms of Figure 7.1

$$\mathcal{E}_0 = \frac{\rho}{\epsilon_0} \quad \text{and} \quad \mathcal{E} - \mathcal{E}_0 = -\frac{P}{\epsilon_0} \, ,$$

i.e. $\epsilon_0 \mathcal{E}_0 = \epsilon_0 \mathcal{E} + P$.

Now the relative permittivity or dielectric constant $\epsilon_r$ is defined as the ratio $\mathcal{E}_0/\mathcal{E}$, there being no free charges at the interface ($\rho = 0$). Thus the relation $\mathcal{E}_0 = \epsilon_r \mathcal{E}$ leads to

$$\epsilon_0 \mathcal{E} + P = \epsilon_0 \epsilon_r \mathcal{E} \, .$$

Whatever additional dielectrics are introduced, $\mathcal{E}_0$ will always be simply $\rho/\epsilon_0$ and therefore the values of both $\epsilon_0 \mathcal{E} + P$ and $\epsilon_0 \epsilon_r \mathcal{E}$ at any point in space are constant, irrespective of the configuration of the dielectric media. This constancy leads to the frequent occurrence of $\epsilon_0 \mathcal{E} + P$ and

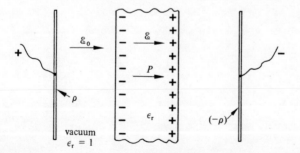

**Figure 7.1.** Dielectric system showing fields, polarisation, and charges.

$\epsilon_0\epsilon_r\mathcal{E}$ in electrostatic work, so much so that it is convenient to replace them by the single symbol $D$.

It is now possible to rewrite the Gauss law in a more useful form. In the equation

$$\int_{\text{surface}} \epsilon_0\mathcal{E}\,\mathrm{d}S = \int_{\text{volume}} (\rho - P)\,\mathrm{d}v$$

we put

$$\int_{\text{volume}} P\,\mathrm{d}v = \int_{\text{surface}} P\,\mathrm{d}S .$$

(This is after all the justification for commonly replacing a uniform volume polarisation by an equivalent surface charge.)  Thence

$$\int_{\text{surface}} (\epsilon_0\mathcal{E} + P)\,\mathrm{d}S = \int_{\text{volume}} \rho\,\mathrm{d}v \quad \text{or} \quad \int_{\text{surface}} D\,\mathrm{d}S = \int_{\text{volume}} \rho\,\mathrm{d}v .$$

The magnitude and frequency dependence of $\epsilon_r$ are of great importance in practice.  The basic properties are apparent on considering the three main contributions to the polarisation.

If any substance whatsoever is subject to an electric field, the electron 'clouds' associated with every atom or molecule will be shifted with respect to the positive nuclei.  The resultant polarisation is quite small and, if there were no other effects, the resultant dielectric constant would exceed unity by less than 1%.  All materials exhibit electronic polarisation and, because of the small mass of the electron clouds, the electronic contribution to $\epsilon_r$ will persist up to optical and even X-ray frequencies.

As light is an electromagnetic radiation, there will be electric interaction between it and any dielectric.  This is summarised by the equation $\epsilon_{re} = n^2$, relating the electronic part of the dielectric constant $\epsilon_{re}$ to the optical refractive index.  The electronic polarisation of any atom (or ion) depends to some extent on the polarisation of adjacent atoms, since their polarisation will produce an electric field additional to the external applied field.  Altering the atomic co-ordinates, by an electric field or by a mechanical strain, will affect the local field relationships, hence altering $\epsilon_{re}$, and varying the optical properties of the dielectric.  This is put to practical use in electro-optic devices for electrically modulating light beams (see Section 7.6).

A larger contribution to $\epsilon_r$ arises if the dielectric molecules are asymmetric, and possess permanent electric dipole moments.  Examples are HCl, $CH_3Br$, $H_2O$, $C_2H_5(NO_2)$, etc.  An external field will tend to align such dipoles and the material becomes polarised.  This alignment is opposed by thermal agitation, and hence $\epsilon_r$ is inversely proportional to the temperature. The comparatively large molecular moments lead to dielectric constants usually well above unity, and in some cases above even a hundred. The greater mass of the molecule results in a lower cut-off frequency, and, dependent on the dielectric and its temperature, this may lie anywhere from a few hundred hertz up to a gigahertz or more (e.g. in $H_2O$ at $20°C$, $f_{co} \sim 10$ GHz, but at $-2°C$, $f_{co} \sim 10$ kHz).

The dipolar contribution relies on the molecules being able to alter their orientation when subject to an external field, and, as a result, it is often suppressed in the solid state. A typical variation of $\epsilon_r$ with temperature is shown in Figure 7.2.

The third contribution is atomic or ionic polarisation, which is the effect of relative atom or ion movements within the crystal lattice. This contribution is often quite small, but in some materials, dependent principally on the crystal structure, ionic polarisation can be of particular importance.

Figure 7.3a shows the structure of a symmetrical ionic solid. It is unpolarised in the normal state. Figure 7.3b shows the effect of applying an external electric field. The centres of positive and negative charge no longer coincide, and the dielectric is polarised. As $\epsilon_r$ is greater than unity, the energy density within the dielectric ($D\mathcal{E}/2 = D^2/2\epsilon_0\epsilon_r$) is less than that outside, and the dielectric expands in the direction of the electric field. This is the phenomenon known as electrostriction. The decrease in electrostatic energy resulting from the expansion is ultimately balanced by the increase in mechanical (elastic) energy.

As the energy density is proportional to the square of the electric field strength, the dielectric expansion is independent of the direction of the electric field. The change in size is therefore proportional only to even powers of $\mathcal{E}$, so that, to a first approximation, electrostriction is a quadratic effect, proportional to $\mathcal{E}^2$. Electrostriction will be especially

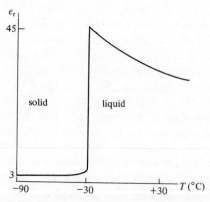

**Figure 7.2.** Variation of $\epsilon_r$ with $T$ for nitromethane.

significant in materials having a high dielectric constant (Mason, 1948) and a large modulus of elasticity.

Figure 7.4a shows an asymmetrical lattice. The charge centres do not coincide, so that the dielectric is already polarised. Normally, this permanent polarisation will be neutralised by the gradual accumulation of surface charges, either from the atmosphere or from an appropriate drift of internal charge carriers.

There are two different bonds involved in the structure of Figure 7.4a, the X bonds and the Y bonds. If the material is stressed mechanically, as in Figure 7.4b, it is very unlikely that the change in orientation and length of the X bonds will exactly balance the change in length of the Y bonds, so far as the resultant polarisation, parallel to the stress, is concerned. The dielectric polarisation therefore changes, and the material is said to be piezoelectric (Brown *et al.*, 1962). Figure 7.4c shows that applying a stress in the opposite direction will alter the apparent polarity of the

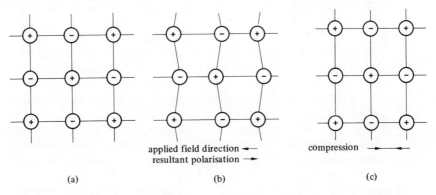

**Figure 7.3.** Centro-symmetric ionic crystal: (a) undisturbed; (b) with electric field applied; (c) in mechanical compression.

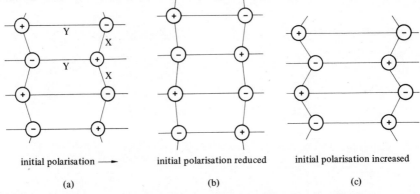

**Figure 7.4.** Piezoelectric crystal structure: (a) undisturbed; (b) in compression; (c) in extension.

piezoelectric charge (provided that the free surface charges do not have time to readjust themselves). To a first approximation, then, piezoelectricity is a linear effect.

Conversely, Figure 7.3c shows the symmetrical dielectric being stressed mechanically. The structural symmetry will not be disturbed and the centres of positive and negative charge remain coincident. This dielectric structure is not piezoelectric.

Placing the piezoelectric material in an electric field will either increase or decrease its length along the applied field, dependent on the relative direction of the field with respect to the dielectric polarisation. The inverse piezoelectric effect is therefore a linear function of the applied field, and may be distinguished from the quadratic effect of electrostriction.

Figure 7.5 shows a ferroelectric material. In small applied fields the ferroelectric behaves like a piezoelectric. However, if larger external fields are applied, the internal structure can 'snap over' to a similar structure but with opposite polarity (Popper, 1956), as in Figure 7.5b. Reducing the field to zero then leaves the ferroelectric in the new state. Plots of polarisation, or electric displacement $D$, therefore, show a 'square-loop' characteristic, as in Figure 7.6, with resultant hysteresis losses in alternating field applications. It is also evident that the small signal values of $\epsilon_r$ will be a function of the d.c. bias field. A typical variation for a ferroelectric material is shown in Figure 7.7.

Ferroelectric characteristics are temperature-dependent for the same reasons as are those associated with ferromagnetism. At a sufficiently high temperature, again called the Curie temperature, ferroelectric alignment will be destroyed. This leads to a typical order-disorder temperature variation in $\epsilon_r$, as, for example, in Figure 7.8. A difference from ferromagnetic behaviour is that at other temperatures, below the Curie temperature, there may be further changes in crystal structure, either altering the ferroelectric properties, or even suppressing them altogether

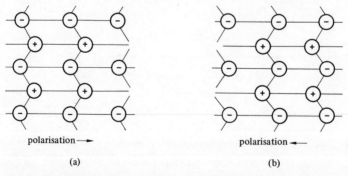

polarisation ⟶                                                          polarisation ⟵

(a)                                                                                      (b)

**Figure 7.5.** Ferroelectric crystal structure (schematic): (a) positive polarisation; (b) negative polarisation.

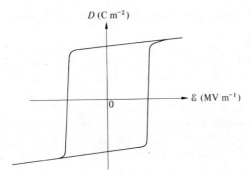

**Figure 7.6.** Single-crystal ferroelectric: $(D, \mathscr{E})$ characteristic.

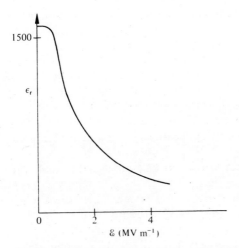

**Figure 7.7.** Variation of $\epsilon_r$ with d.c. bias field for single-crystal BaTiO$_3$.

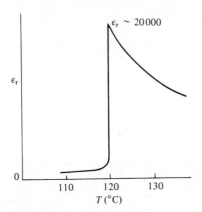

**Figure 7.8.** Variation of $\epsilon_r$ with $T$ for single-crystal BaTiO$_3$.

156

Chapter 7

(Devonshire, 1949). Thus Rochelle salt is ferroelectric only between −18°C and +24°C.

In practical applications, barium titanate is usually prepared as a ceramic material, comprising a mass of individual crystallites. These are roughly aligned along any desired axis by applying a suitable electric field to the ceramic as it is cooled from above the Curie temperature. The resultant material is mechanically hard and can be made in larger and more complex shapes than would be possible with single crystals. The most important change in ferroelectric properties is that the square-loop characteristic is lost, as shown in Figure 7.9. In fact, the reduction in loop area renders the material more suitable for high-frequency use, since hysteresis heating is proportional to the loop area. At the same time, the dielectric constant and saturation polarisation are somewhat reduced.

The temperature dependence of ferroelectric parameters is not ideal for many practical purposes (see, for example, Figure 7.8) and considerable effort has been devoted towards shifting the $\epsilon_r$ peak towards more usual working temperatures, and to broadening it out. A very promising approach has been to alter the composition, using various combinations and ratios of, for example, Pb, Sr, Ba with $TiO_3$, $ZrO_3$, etc. Typical results are shown in Figure 7.10 (Vincent, 1951).

Aligned ferroelectric ceramics are, of course, piezoelectric, and a very popular material for transducer applications is ceramic lead zirconium titanate (PZT).

It may be seen, then, that ferroelectric materials are suitable for use in at least three different situations; they can be used simply as high-$\epsilon_r$ dielectrics, as square-loop materials, or as effective piezoelectric materials.

A most important feature that leads to the practical exploitation of various different dielectric phenomena is the tremendously wide range of dielectrics available. This contrasts markedly with the applications of

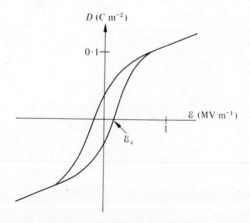

**Figure 7.9.** Ceramic ferroelectric: $(D, \mathcal{E})$ characteristic.

magnetic phenomena where the only materials with permeabilities significantly greater than unity almost always involve the three elements iron, cobalt, and nickel. Any new material with a high electrical resistivity is potentially a useful dielectric and may be found to have superior properties in some way or another to dielectrics already in existence.

In fact, the range of dielectrics available, and the likelihood of more than one characteristic being significant in any specific application, make it difficult to be sure of choosing the most suitable material for a particular task. In many situations either the electric breakdown field strength $\mathcal{E}_B$ or the conductivity $\sigma$ is at least as important as $\epsilon_r$, and there is generally no way of predicting what these characteristics will be in a new dielectric, and how they will be related to one another. A high value of $\epsilon_r$ often means a low value of $\mathcal{E}_B$, but usually the relationship cannot be put in a more specific form.

Comparison of different dielectrics is often simplified by the concept of a material having various figures of merit. The exact formulation of the figure of merit depends on the application envisaged. Some of the following sections will show the relevance of this approach.

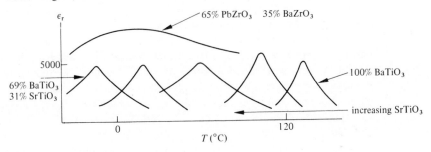

**Figure 7.10.** Variation of $\epsilon_r$ with $T$ for ceramic ferroelectrics of different compositions.

## 7.2 Fixed capacitors

Capacitor dielectrics are prepared either as very thin films, for example by electrolytic action, or as substantially thicker layers made from bulk material, for instance by slicing. The different ranges of thickness considerably influence the choice of dielectric material, and the characteristics of the finished capacitor.

The electrolytic capacitor (Burger and Young, 1963) relies on an oxide film as dielectric, commonly with aluminium or tantalum as the anode. The oxide film is formed by electrolysis; a voltage somewhat in excess of the intended capacitor working voltage is applied from a current-limiting circuit. As the oxide thickness increases, so does its actual resistance, until eventually the current drops to a very low value. The capacitor is then suitable for use, and the leakage current will be reasonably small, provided that the applied voltage is kept less than the formation voltage.

Originally, the electrolytic capacitor was used with a liquid electrolyte still in the 'can', but obvious disadvantages led to this wet electrolyte being replaced by a damp paste. The solid electrolytic capacitor is a fairly recent development. The oxide film is formed essentially as above, but the wet electrolyte is then completely removed and the embryo capacitor thoroughly dried. The surface of the oxide film is then coated with a conducting layer, and the capacitor completed. The solid electrolytic capacitor has the advantage that, being dry, it may be used over considerably wider temperature ranges, e.g. from $-50°C$ to $+120°C$ compared with $-35°C$ to $+65°C$ for the older wet or 'dry' electrolytic capacitors. Since the final capacitance is proportional to the electrode area, the anode is usually in the form of etched foil, or sponge-like material, so that the effective surface area is considerably increased.

The oxide film is typically between $10^{-9}$ m and $10^{-7}$ m thick, so that its volume is negligible in comparison with that of the anode. This means that the effective surface area is proportional to the overall volume of the capacitor. If the dielectric oxide thickness is $t$, and the effective surface area is $A$, then the capacitance $C$ will be $\epsilon_0\epsilon_r A/t$. The maximum working voltage $V_B$ depends on the dielectric breakdown field strength $\mathcal{E}_B$; $V_B = \mathcal{E}_B t$. Eliminating $t$ gives $CV_B = \epsilon_0\epsilon_r A \mathcal{E}_B$ which, for a given capacitor volume, is proportional to $\epsilon_r\mathcal{E}_B$.

Thus the product $CV_B$ is constant for a given volume and the material figure of merit is $\epsilon_r\mathcal{E}_B$. All other things being equal, a capacitor dielectric should be chosen with a high value of $\epsilon_r\mathcal{E}_B$. Aluminium and tantalum are popular anode materials for electrolytic capacitors, not because their oxides have specially high values of $\epsilon_r$, but because $\mathcal{E}_B$ is very high. For example, for tantalum oxide, $\epsilon_r \sim 28$ and $\mathcal{E}_B \sim 10^8$ V m$^{-1}$, whence the material figure of merit is about $3 \times 10^9$ V m$^{-1}$.

Electrolytic capacitors that are commonly available range from 1 $\mu$F to $10^5$ $\mu$F in capacitance, with breakdown voltages from 500 V down to just 3 or 4 V. Because of the small leakage current, they tend to be rather lossy, with $\tan\delta$ around $0\cdot005$.

For low-loss capacitors, dielectric layers are usually prepared by mechanical means from bulk material, and even the thinnest are very much thicker than the oxide films found in electrolytic capacitors. To get a reasonable capacitance it is, therefore, necessary to use a material with a high value of $\epsilon_r$, and, indeed, the ferroelectric ceramics often employed have $\epsilon_r$ values of some thousands (Herbert, 1965). Nevertheless, $\mathcal{E}_B$ is again important, in fact even more so, as the material figure of merit will be shown to equal $\epsilon_r\mathcal{E}_B^2$.

In these capacitors, the dielectric is usually mechanically self-supporting, and the metal electrodes may be deposited on to the dielectric in very thin layers. Under these circumstances the dielectric accounts for almost the entire volume of the capacitor, and this is responsible for the different formulation of the material figure of merit.

Suppose such a capacitor is built having a volume $v$ as in Figure 7.11

where the electrodes are negligible in thickness compared with the dielectric sheets. Then the number of sheets is given by $n = h/t$, and the total capacitance is just

$$C = \frac{n\epsilon_0\epsilon_r A}{t} = \frac{hA\epsilon_0\epsilon_r}{t^2} = \frac{v\epsilon_0\epsilon_r}{t^2} \; .$$

The breakdown voltage is again $V_B = \mathcal{E}_B t$, whence, on eliminating $t$, $CV_B^2 = \epsilon_0\epsilon_r\mathcal{E}_B^2$ per unit volume. With this construction then, the product $CV_B^2$ is constant for a given volume, and the material figure of merit is $\epsilon_r\mathcal{E}_B^2$.

Although material with high $\epsilon_r$ may be used, the capacitance is much less than for electrolytic capacitors, because of the greater dielectric thickness. Capacitance values do not usually exceed 1 $\mu$F but breakdown voltages can be made very high by suitable choice of dielectric and geometry. The losses through dielectric conduction are usually quite negligible, and $\tan\delta$ is less than $0\cdot0001$ (for ferroelectric ceramics).

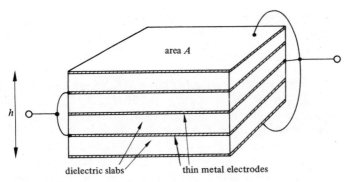

**Figure 7.11.** Construction of parallel-plate slab capacitor.

### 7.3 Piezoelectric transducers

Piezoelectric material may be used as coupling between mechanical and electrical systems. The simplest transducer would be a block of piezoelectric material with conducting electrodes on two opposite faces, as in Figure 7.12. Application of an external voltage to the electrode connections results in a mechanical stress within the transducer, and conversely an applied mechanical strain results in a change in the internal electric field.

For moderate stresses and strains, the transducer behaves as a linear two-port element, relating $D$ and $\mathcal{E}$ to $S$ (strain) and $T$ (stress). The characteristics may therefore be put in the form

$$D = aT + b\mathcal{E} \quad \text{and} \quad S = cT + d\mathcal{E} \; ,$$

$a$, $b$, $c$, and $d$ being constants of the material. From these equations it can be seen that $b = (\partial D/\partial\mathcal{E})_T$, which is usually written $\epsilon^T$, and $c = (\partial S/\partial T)_\mathcal{E}$, usually written as $s^E$. (Note that $\epsilon^T$ will be the usual dielectric constant $\epsilon_0\epsilon_r$ provided that the method of measuring $\epsilon_r$ incurred no mechanical stressing of the material.)

Also the total internal energy $U$ is a function of $D$, $\mathcal{E}$, $S$, and $T$, such that

$dU = T\,dS + \mathcal{E}\,dD + dQ$, where $dQ$ is any heat transferred.

For many applications $dQ$ may be neglected, whence

$dU = T\,dS + \mathcal{E}\,dD$ .

Now $dU$ is an exact differential, so that

$$\left(\frac{\partial \mathcal{E}}{\partial S}\right)_T = \left(\frac{\partial T}{\partial D}\right)_{\mathcal{E}}$$

or $a = d$. The original equations may thus be rewritten in their usual forms,

$S = s^E T + d\mathcal{E}$

and

$D = dT + \epsilon^T \mathcal{E}$ .

Since transducers are often used for conversion of mechanical to electric energy, it is useful to define a coupling coefficient relating the proportion of energy input in one form to that which may be extracted in the other.

Suppose a mechanical stress $T$ is applied to the transducer. This will require mechanical energy input of $s^E T^2/2$. The mechanical strain resulting from $T$ produces a piezoelectric stress $\mathcal{E}$, so there is electrical energy involved. The electrical energy is $D\mathcal{E}/2$ or $\epsilon_0\epsilon_r\mathcal{E}^2/2$ per unit volume.

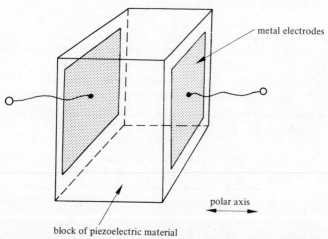

**Figure 7.12.** Simple piezoelectric transducer.

Now $D = dT + \epsilon^T \mathcal{E}$, and, if no electrical energy is actually extracted, there can be no charge movement and $D$ remains zero. Thus, $\mathcal{E} = -dT/\epsilon^T$, whence the ratio of available electrical energy to mechanical energy input equals $d^2/\epsilon^T s^E$. The same result would be found by considering an applied electric stress ($\mathcal{E}$) and evaluating the ratio of mechanical energy available to electrical energy input. The coupling coefficient $k$ of a piezoelectric material is therefore defined by $k^2 = d^2/s^E \epsilon^T$. (This may be compared to the electrical system shown in Figure 7.13. The equations relating stresses to strains are $V_1 = j\omega L_1 I_1 + j\omega M I_2$ and $V_2 = j\omega M I_1 + j\omega L_2 I_2$, where it is well known that the coupling coefficient is given by $k^2 = M^2/L_1 L_2$.)

Many applications of piezoelectric transducers involve a one-way conversion of energy, from electrical to mechanical or *vice versa*. There are also applications in which the mechanical characteristics are reflected back into the electrical side to produce electrical components having valuable practical features, and in which the mechanical side of the transducer is of no direct interest to the user. Various examples of transducer applications follow.

A common example of mechanical-to-electrical conversion is the crystal pick-up used in many record players. The mechanical movement of the stylus results in an output about one volt in amplitude. Pure crystal transducers are limited to fairly low electrical outputs as they are liable to fracture in high stresses. However, the advent of ferroelectric ceramics, with high resistance to mechanical crushing has led to transducers capable of generating tens of kilovolts. The so-called spark pump is shown in Figure 7.14. Hand pressure applied via mechanical leverage is sufficient to produce sparks about 5 mm long in air (Pearcy, 1965). Applications of high-voltage transducers generally lie in the ignition of inflammable gases, for example in domestic gas fires or in petrol engines. High coupling coefficients are desirable here, but at the same time it is essential that a high electric field strength be developed, otherwise the terminal voltage will be insufficient to produce a spark. Now $D = dT + \epsilon^T \mathcal{E}$. Therefore, as the electrical circuit is open until the spark occurs, $D = 0$ and $\mathcal{E} = -dT/\epsilon^T$. Thus, it is important in these applications to choose a material with a high coupling coefficient $k$ and a high value of $d/\epsilon^T$.

With the successful production of materials having high coupling coefficients, it has become possible to convert electrical energy to

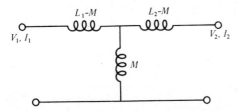

**Figure 7.13.** Equivalent circuit of lossless transformer.

mechanical at frequencies from d.c. up to hundreds of kilohertz. The efficiency of the higher-frequency transducers led to the rapid growth of interest in ultrasonics (Brockelsby, 1963), whereas the low-frequency transducers have led to the conversion of electrical energy to rotary, or large-amplitude linear, movements. By cementing two oppositely polarised slabs of material together, considerable amplification of the mechanical effect occurs (this is similar to the effect of heating a bimetallic strip). Figure 7.15 shows two forms of 'bimorph', one as described above (Orwell,

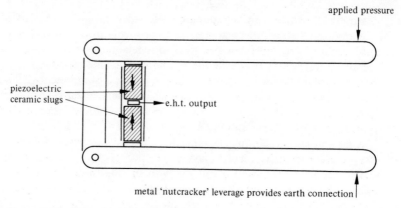

**Figure 7.14.** Piezoelectric spark generator.

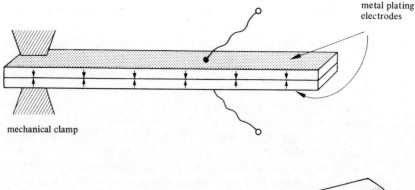

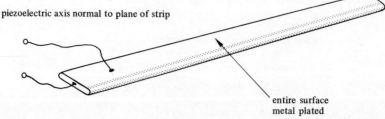

**Figure 7.15.** Two forms of piezoelectric bimorph.

1963), and the other, which is simpler to manufacture, comprising a uniformly polarised ferroelectric with a longitudinal hole for inserting one of the wire electrodes. It will be appreciated that, in the latter form of bimorph, the electric field in the upper half of the material will be in the opposite direction to that of the lower. Two applications are shown in Figure 7.16: the mechanical resonant frequency in each is adjusted to be equal to that of the a.c. voltage. In Figure 7.16b the vibratory motion is converted to rotation and may be used, for instance, to drive a clock mechanism.

If the mechanical side of an electrical-to-mechanical transducer is comparatively lightly loaded, then the strain will be significant and the stress minimal. The crystal is in effect mechanically short circuited. Under these conditions $S = d\mathcal{E}$ and, in applications where large mechanical movement is desired with very little resistance (e.g. as in the clock and bell mentioned above), a material must be chosen with a high value of $d$.

One of the best-known transducer applications must surely be the use of quartz crystals in precision electronic oscillators. The mechanical resonance properties of the quartz are reflected in the electrical characteristics,

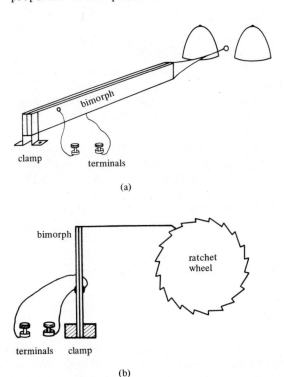

(a)

(b)

**Figure 7.16.** Application of low-frequency transducers: (a) electric bell; (b) rotary motion.

yielding an impedance that varies with frequency somewhat as in an orthodox parallel LCR circuit (Mason, 1934). However, there are significant differences. The intrinsic $Q$ factor of quartz is far higher than that of an electrical LCR circuit; for the latter, $Q$ might be some tens, or even hundreds with careful design, whereas, for quartz, $Q$ is measured in tens of thousands, or even hundreds of thousands. Furthermore, by slicing the quartz crystal appropriately, the temperature coefficient may be made extremely small; for example, with a so-called GT cut the resonance frequency will not vary by more than two or three parts per million over the temperature range 0° to 100°C. This also is far superior to the LCR circuit where temperature changes affect physical dimensions and hence component values. An apparent shortcoming of quartz is the low value of coupling coefficient, $k$ being typically between 0·09 and 0·14. However, the loose coupling effectively isolates the mechanical resonance from interference by the external electrical circuit. This is a useful feature in precision oscillators as it prevents $Q$ being inadvertently lowered by excessive loading of the electrical side. Another effect of the high $Q$ and loose coupling is that it is not possible to vary the resonant frequency by electrical methods by more than a few parts per million. The resonant frequency is determined during the crystal manufacture, so that careful grinding, and controlled deposition of the electrodes (which are, of course, directly coupled to the resonance), are essential.

The frequency range covered by quartz crystals depends on their size, and on the mode of vibration. Commonly, quartz crystal oscillators may be built to operate between some tens of kilohertz to tens of megahertz, and the resonant frequency can be stable to a few parts in $10^9$ (Spencer-Jones, 1945).

The low mechanical losses in quartz lead to its use in electromechanical delay lines (Gibson, 1965). An electrical signal is converted into a mechanical wave which travels (at the speed of sound) through the quartz to a detector where it is reconverted to an electrical signal. To accommodate longer delays, the wave can be made to reflect back and forth within the quartz in the manner shown in Figure 7.17. The wave velocity is about 4 km s$^{-1}$ so that total path lengths of some metres lead to delays of the order of a millisecond, and are suitable for signals up to about 100 MHz. Quartz is commonly used here because of its low transmission losses.

Other systems working on an electrical-to-mechanical-to-electrical basis make use of crystals with two sets of electrodes. Such elements behave very much as electrical transformers, with a 'turns ratio' dependent on the geometries of the electrodes and crystal. Piezoelectric material having a high value of $k$ is chosen and the mechanical losses adjusted during the material preparation to give the desired $Q$ and bandwidth. Satisfactory i.f. transformers have been manufactured for use in domestic radio receivers (Crawford, 1961). Yet again, by building in a high 'turns ratio', it is possible to generate kilovolts or more, from, for example, a 12 V battery. The voltage step-up is emphasised by operating at the resonant frequency of the piezoelectric material and the overall size is very much less than that of conventional e.h.t. transformers.

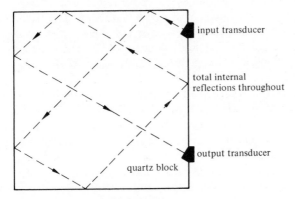

input transducer

total internal
reflections throughout

output transducer

quartz block

Figure 7.17. Multi-reflection quartz delay line (plan view).

## 7.4 Ferroelectric applications

The applications considered here rely specifically on the loop in the ferroelectric $(D, \mathcal{E})$ characteristic, and the possibility of reversing $D$ if a critical reverse field strength $\mathcal{E}_c$ is exceeded. Ferroelectric materials are also used as high $\epsilon_r$ dielectrics in capacitors, as piezoelectric elements in transducers, and as non-linear dielectrics in dielectric amplifiers. These applications are referred to elsewhere in this chapter rather than here since, in them, the ferroelectric aspect as such is not of direct concern.

If ferroelectric material is placed in an alternating electric field, whose amplitude exceeds the coercivity $\mathcal{E}_c$, then heat will be generated owing to the hysteresis loss. The temperature of the material will therefore rise. As the temperature rises, the $(D, \mathcal{E})$ characteristic shrinks. In fact, the shrinkage is quite small to within a few degrees of the Curie temperature. Within the last few degrees below $T_c$ the loop shrinks virtually to nothing, and the hysteresis heat generation falls to zero. Figure 7.18a shows the variation in heat generation as the temperature rises. The subsidiary curves, Figure 7.18b, show varying heat losses from the crystal through conduction, etc., in different ambient temperatures. In each case, the temperature of the material will settle at the intersection of curve (a) with the appropriate curve (b). Despite considerable variation in ambient temperature, the temperature of the ferroelectric material stabilises just below $T_c$.

If the amplitude of the drive field is less than $\mathcal{E}_c$ at ambient temperature, the heat generated will be small. However, $\mathcal{E}_c$ decreases as $T$

rises and, at an elevated temperature, the drive field will exceed $\mathcal{E}_c$ so that significant hysteresis heating occurs. Figure 7.18c shows a heat generation curve with a small a.c. field. If the ambient temperature is sufficiently high, temperature stabilisation will occur (point X in Figure 7.18) but, with a lower ambient temperature, the crystal temperature will merely rise a little with no stabilisation (point Y in Figure 7.18). In practice, then, there is no difficulty in deciding whether or not the a.c. field is large enough for temperature stabilisation to occur. If the field strength is inadequate, the ferroelectric temperature barely gets 'off the ground'.

A ferroelectric device designed to temperature stabilise in this way is called a TANDEL (Malek *et al.*, 1964; Glanc, 1964) [1]. Such a device is inherently suitable for use in frequency multipliers and dielectric amplifiers where the effects of changes in ambient temperature must be minimised.

In the absence of external electric fields, a ferroelectric material possesses two stable states, dependent on the direction of polarisation with respect to the ferroelectric axis. As with a ferromagnetic material, this has led to its use in computer memory systems (Prutton, 1959).

A basic one-bit store is shown in Figure 7.19. A negative pulse applied to Z will polarise the ferroelectric appropriately and a subsequent positive pulse will reverse the polarisation, producing a positive output signal across the resistor. In practice, positive pulses at Z could be used to write in 'zeros' and negative pulses for 'ones'. A positive pulse can also serve as a 'read' signal producing a positive output signal if a 'one' had been in store. This again (cf. the magnetic store) is a destructive system, since the reading leaves all the bits in the same (zero) state, and rewriting may be necessary. A simple selection system for a multi-bit store is shown in

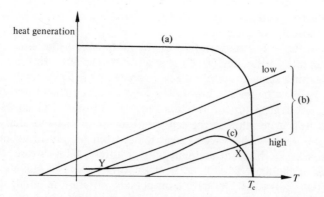

**Figure 7.18.** Thermal considerations in an a.c. driven ferroelectric: (a) heat generation; (b) heat losses at various ambient temperatures; (c) heat generation with small drive field.

[1] Temperature Autostabilising Non-linear Dielectric ELement.

Figure 7.20. The appropriate x conductor is earthed via the read-out circuit, and the y conductor connected to a potential that is sufficient to switch the ferroelectric material into the intended state. Unaddressed elements should in principle be undisturbed, but capacitive coupling effects result in stray pulse fields which, despite perhaps being very small, can seriously affect the stored information.

Ferroelectric memories have not achieved widespread popularity, mainly because of material difficulties. Barium titanate is the only ferroelectric seriously considered for computer applications, and, even so, it has two severe shortcomings relating to its coercivity. Although Figure 7.6 represents the characteristics over a complete cycle of the applied field, it is found that repeated application of reverse pulses of amplitude $\mathcal{E}_r$, where $\mathcal{E}_r < \mathcal{E}_c$ will eventually reverse the ferroelectric polarisation. The number of reverse pulses necessary to do this tends to be infinite only as $\mathcal{E}_r$ tends to zero. Clearly, then, unaddressed memory elements on selected x and y

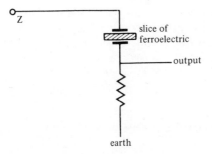

**Figure 7.19.** Basic ferroelectric memory element.

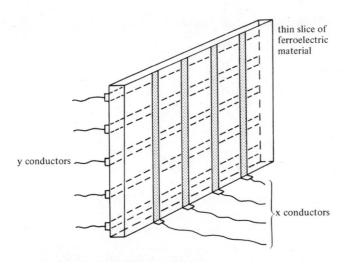

**Figure 7.20.** Ferroelectric x-y selection.

conductors can be inadvertently disturbed and their stored information lost. The remedy is to rewrite the entire memory at suitable intervals, but this is obviously a design complication. The other difficulty is that $\mathscr{E}_c$ is found to increase with the effective cycling speed of the ferroelectric. Thus, as the repetition rate increases, the heat generated within the ferroelectric rises very rapidly, and the temperature may rise above the Curie temperature. Ferroelectric memories are limited because of this to cycle times of at least several microseconds, which, in general, is too long for modern computers.

### 7.5 Dielectric amplifiers

In some materials, particularly the ferroelectric ceramics referred to earlier, $\epsilon_r$ is a non-linear function of electric field strength. Any capacitor made with such a material as its dielectric will therefore exhibit a non-linear variation in capacitance with applied voltage.

A practical application of any voltage-dependent capacitor is to be found in dielectric amplifiers (Vincent, 1951). These provide power gain rather than voltage gain, and, in fact, considerable voltage attenuation may occur. However, one particular feature of dielectric amplifiers is their very high input resistance at low frequencies (quasi d.c.).

Consider the circuit shown in Figure 7.21a, where $C_1$ is a voltage-

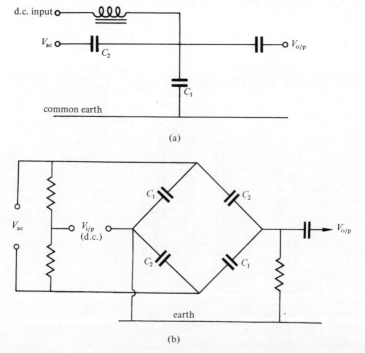

(a)

(b)

**Figure 7.21.** Basic circuits of dielectric amplifiers (all $C_1$ are non-linear capacitors).

dependent capacitor. The output voltage is simply $V_{o/p} = V_{ac}C_1/(C_1+C_2)$. Variation in the d.c. bias (signal) will change $C_1$ and, hence, the amplitude of $V_{o/p}$. A bridge circuit is shown in Figure 7.21b. With $V_{i/p}$ equal to zero the bridge is balanced and there is no voltage, alternating or otherwise, across the output resistor. As $V_{i/p}$ varies, the bridge becomes unbalanced, and an a.c. output appears. This may be rectified as necessary, to produce a final d.c. output voltage that, with suitably chosen dielectric, bias, etc., may be a reasonably linear function of $V_{i/p}$.

An alternative approach is to use the non-linear capacitor as part of a high-$Q$ tuned circuit. Small variations in capacitance can then cause very large changes in the impedance, and high amplification has been achieved in this way.

One of the main attractions of dielectric amplifiers is the almost infinite input impedance, due to the input terminal being connected only to capacitors. For example, a dielectric amplifier with an input resistance of $10^{12} \, \Omega$ has been described (Ettinger, 1966).

As the input equivalent circuit is essentially just a capacitor $C_{i/p}$ the amplifier may be calibrated as a charge amplifier. At the input terminal $C_{i/p}V_{i/p} = Q_{i/p}$, $Q_{i/p}$ being the charge associated with an input voltage $V_{i/p}$. The output voltage $V_{o/p}$ is proportional to $V_{i/p}$, whence $V_{o/p}$ is also proportional to $Q_{i/p}$. By making the input capacitance small, a high sensitivity $dV_{o/p}/dQ_{i/p}$ can be achieved.

## 7.6 Electro-optic devices

Many of the proposed uses of lasers depend on being able to modulate or to deflect the emergent beam of high frequencies, preferably at some gigahertz. One promising way of achieving this is by using electro-optic devices. In dielectrics with axial symmetry, the electronic contribution $\epsilon_{re}$ to the dielectric constant varies with the angle between that axis and the electric field involved. As $\epsilon_{re} = n^2$, electromagnetic waves of different polarisation will generally experience different refractive indices.

One of the first successful electro-optical devices was the Kerr cell using nitrobenzene as dielectric (Dunnington, 1931). Nitrobenzene is a dipolar liquid and an electric field bias can produce sufficient molecular alignment for detectable differences in $\epsilon_{re}$. A similar, although smaller, effect is observed in non-polar liquids where the constituent molecules have an axis of symmetry. In present-day Kerr cells intended for microwave applications, non-polar carbon disulphide may be employed (Holshouser et al., 1961). Although its electro-optic coefficient is only about 1/150 times that of nitrobenzene, the use of nitrobenzene at gigahertz frequencies is entirely precluded by an excessively high loss factor.

Figure 7.22 shows a Kerr cell being used to modulate a light beam. The incident light is polarised at 45° to $\mathcal{E}_0$ and may be considered to comprise two plane polarised waves, one polarised parallel to $\mathcal{E}_0$, and the other

normal to $\mathscr{E}_0$. These waves experience slightly different refractive indices
on passing through the liquid and the plane of polarisation rotates. The
light intensity emergent from the analyser is thus dependent on the
amplitude of $\mathscr{E}_0$.

The above electro-optic effect is quadratic in terms of $\mathscr{E}_0$, since it makes
no difference to the light beam whether the molecular alignment normal
to the beam is in one direction rather than any other. The effect is
immediately suitable for use in on-off electro-optic shutters, but for linear
modulation the cell must be biassed electrically and only a comparatively
small modulation superimposed.

The use of liquid dielectrics in electro-optic devices does have practical
disadvantages. It is important to maintain absolute purity, and heat
generation can lead to the build-up of high pressures within sealed cells.
Not surprisingly solid-state alternatives are in demand.

Clearly what is needed is a crystal structure readily distorted by
application of electric fields, and, as might be expected, the most
successful materials are well known piezoelectrics and ferroelectrics,
e.g. $KH_2PO_4$ (KDP), $BaTiO_3$, etc. Again, as might be expected, the
electro-optic effect is linear in terms of the applied field if the field is
applied along the polar axis.

There are in fact two contributions to the linear effect. One is a
consequence of the piezoelectric strain induced by the applied field
whereas the other is a direct effect. At gigahertz frequencies there is no
possibility of any mechanical response and only the direct effect can be
operative.

To distinguish devices working in the linear mode from those operating
in the quadratic regime, the former are often referred to as Pockels cells
and the latter as Kerr cells.

The use of electro-optic devices for modulating and deflecting light
beams is, in principle at least, fairly straightforward (Chen et al., 1966).
Modulation is usually achieved by passing plane-polarised light through the
dielectric with the plane of polarisation suitably oriented with respect to

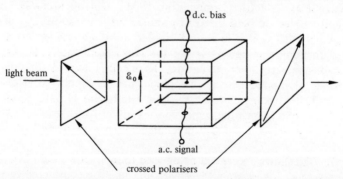

**Figure 7.22.** Kerr cell biassed for linear modulation.

the electric field, and crystal axis. Beam deflection can be achieved with an electro-optic prism, as shown in Figure 7.23. Varying the electric field varies the refractive index of the prism, and so changes the angle of deflection. 'Straight-through' devices include a second prism of similar refractive index, but which is not electro-optically active.

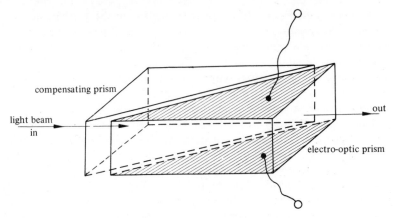

**Figure 7.23.** 'Straight-through' electro-optic beam deflector.

**References**
Brockelsby, C. F., 1963, *J. Sci. Instr.*, **40**, 153.
Brown, C. S., Kell, R. C., Taylor, R., and Thomas, L. A., 1962, *Proc. Inst. Elec. Engrs. (London), Pt.B*, **109**, 99.
Burger, F. J., and Young, L., 1963, *Progr. Dielectrics*, **5**, 1.
Chen, F. S., Gensic, J. E., Kurtz, S. K., Skinner, J. G., and Wemple, S. H., 1966, *J. Appl. Phys.*, **37**, 388.
Crawford, A. E., 1961, *J. Brit. Inst. Radio Engrs.*, **21**, 353.
Devonshire, A. F., 1949, *Phil. Mag.*, **40**, 1040.
Dunnington, F. G., 1931, *Phys. Rev.*, **38**, 1506.
Ettinger, G., 1966, *Electron. Eng.*, **38**, 731.
Gibson, R. W., 1965, *Ultrasonics*, **3**, 49.
Glanc, A., 1964, *Solid-State Electron.*, **7**, 671.
Herbert, J. M., 1965, *Proc. Inst. Elec. Engrs. (London), Pt.B*, **112**, 1474.
Holshouser, D. F., Van Foerster, H., and Clark, G. L., 1961, *J. Opt. Soc. Am.*, **51**, 1360.
Malek, Z., Mastner, J., Hrdlicka, J., and Janta, J., 1964, *Solid-State Electron.*, **7**, 655.
Mason, W. P., 1934, *Bell System Tech. J.*, **7**, 20.
Mason, W. P., 1948, *Phys. Rev.*, **74**, 1134.
Orwell, R. J. F., 1963, *Ultrasonics*, **1**, 49.
Pearcy, C., 1965, *Electron. Eng.*, **37**, 656.
Popper, P., 1956, *J. Inst. Elec. Engrs. (London)*, **2**, 450.
Prutton, M., 1959, *J. Brit. Inst. Radio Engrs.*, **19**, 93.
Spencer-Jones, H., 1945, *Endeavour*, **4**, number 16.
Vincent, A., 1951, *Electronics*, **24**, number 12, 84.

# Superconductors

## 8.1 Applications of type I superconductivity

The most obvious characteristic of a superconductor is its ability to carry an electric current with no voltage drop. This suggests that cables made from such a material could effect great savings in the transmission of power, even allowing for the fact that they would have to be cooled to very low temperatures (superconductivity has not been observed at temperatures greater than about $20°K$). This certainly is a possible application, but it is not so simple to put into practice as it sounds. The difficulty lies in the fact that a superconductor becomes normal when the current exceeds a limiting value.

The origin of this critical current is revealed if we consider one of the most important features of superconductivity, namely the Meissner effect. Meissner and Ochsenfeld (1933) first showed that the magnetic field does not penetrate appreciably into an ideal superconductor; actually the field decays to zero at a depth of about 1000 Å below the surface. In other words, a superconductor may be viewed as an almost perfect diamagnetic material. The magnetic energy of a superconductor in a field $B$ is greater by $B^2/2\mu_0$ per unit volume than it would be in a normal metal (if we disregard the weak paramagnetic susceptibility of the latter). In zero magnetic field, the superconducting state is preferred to the normal state because the superconducting free energy $G_S(0)$ is lower than the normal free energy $G_N(0)$. However, when the magnetic field is applied, the Meissner effect causes the energy difference to become less than $G_N(0) - G_S(0)$. The energies of the two states become equal when the field reaches a value $B_c$ given by

$$\frac{B_c^2}{2\mu_0} = G_N(0) - G_S(0) .$$ 

(8.1)

At higher magnetic fields the material prefers the normal state; the field $B_c$ is known as the critical field.

Silsbee pointed out that, when a current is passed through a superconducting wire, a magnetic field is produced and he showed that the critical current is the value at which the magnetic field at the surface reaches the critical value $B_c$. That is

$$\frac{I_c}{2\pi a} = \frac{B_c}{\mu_0}$$ 

(8.2)

where $I_c$ is the critical current for a wire of radius $a$. Unfortunately the values of $B_c$ are rather low for all type I superconductors, even at $0°K$. Table 8.1 gives values of the critical field at the absolute zero and the critical temperature of most of the elemental superconductors.

**Table 8.1.** Critical temperatures and critical fields at $0°K$ for elemental superconductors.

| Element | $T_c$ (°K) | $B_c(0)$ (T) |
|---------|-----------|--------------|
| Al | 1·20 | 0·010 |
| Cd | 0·56 | 0·003 |
| Ga | 1·09 | 0·005 |
| Hg ($\alpha$) | 4·15 | 0·041 |
| Hg ($\beta$) | 3·95 | 0·034 |
| In | 3·40 | 0·029 |
| La ($\beta$) | 5·95 | 0·160 |
| Mo | 0·92 | 0·010 |
| Nb | 9·25 | 0·195 |
| Os | 0·65 | 0·007 |
| Pb | 7·19 | 0·080 |
| Re | 1·70 | 0·020 |
| Ru | 0·49 | 0·007 |
| Sn | 3·74 | 0·030 |
| Ta | 4·48 | 0·078 |
| Th | 1·37 | 0·016 |
| Ti | 0·39 | 0·010 |
| Tl | 2·36 | 0·017 |
| V | 5·30 | 0·131 |
| Zn | 0·91 | 0·005 |
| Zr | 0·55 | 0·005 |

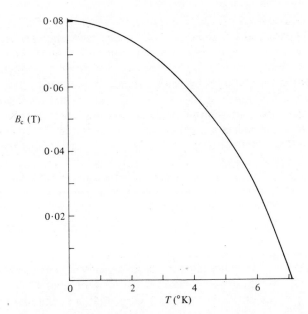

**Figure 8.1.** Critical field plotted against temperature for lead.

As the temperature rises above $0°K$, the critical field falls, reaching zero at the temperature $T_c$. It was found empirically that

$$\frac{B_c}{B_c(0)} \approx 1 - \left(\frac{T}{T_c}\right)^2 \tag{8.3}$$

where $B_c(0)$ is the critical field at $0°K$. Figure 8.1 shows how the critical field varies with temperature for one of the most commonly studied superconductors, lead. Thus, if we were to make a cable from Pb wire, the critical field being about $0·056$ T at $4·2°K$ (liquid helium temperature), the critical current would be about 140 A per mm diameter. This suggests that a considerable quantity of superconductor would be needed if reasonably large currents were to be carried any distance. However, the current in the superconductor is confined to the surface regions; so it is really necessary to have only a thin cylindrical shell of the required diameter.

As we shall see later, much higher critical fields can be reached with type II superconductors, but these often give critical currents that are much lower than Silsbee's rule predicts. Moreover, only a direct current is carried without loss in a type II superconductor. Type I superconductors are perfectly loss free for alternating current at ordinary frequencies, as well as direct current.

### 8.1.1 Zero resistance

We shall now consider the possibilities of using the zero-resistance phenomenon in specific devices.

One of the most obvious uses is to improve the sensitivity of a *galvanometer* (Pippard and Pullan, 1952). Ordinary galvanometers are limited in their sensitivity by the resistance of their coils. By making a coil from a superconducting wire, a given voltage will produce a much larger current that is limited only by the resistance of the leads. The increase of sensitivity is gained at the expense of a rise in the response time, since the time constant is given by the ratio $L/R$, where $L$ is the inductance and $R$ is the resistance. Such an instrument can be sensitive to about $10^{-12}$ V, but another superconducting galvanometer utilising the Josephson effect, which will be described later, has better characteristics.

More practicable has been the use of superconductors to make *high-Q cavities*. For example, a $Q$ of $4 \times 10^8$ has been obtained in an unloaded rectangular cavity made from lead and operated at $1·8°K$ (Fairbank *et al.*, 1963). The upper frequency limit of a superconducting cavity is set by the losses that occur when $h\omega$ for the electromagnetic radiation ($\omega$ being the angular frequency) is equal to the superconducting energy gap (Bardeen *et al.*, 1957). This gap is temperature-dependent, but generally of the order of $kT_c$, except at temperatures that are very close to $T_c$. The limiting frequency turns out to be near to 1 GHz, that is in the low microwave region.

### 8.1.2 Change of resistance near the critical temperature

It is usually observed that the transition between the normal and super-conducting states occurs over a small but finite range of temperature. The complete change may take place within about $10^{-2}$ degK for a rather pure and perfect crystal, while for other materials it might be spaced over one degree Kelvin or more. Figure 8.2 shows the behaviour that has been observed for various samples of tin.

A superconducting wire obviously serves as a very sensitive *thermometer* over the temperature range of the transition, but the narrowness of this range and a dependence of the resistance on magnetic field make the device rather impracticable. A superconducting *switch* has proved to be rather more useful (Thomas, 1965).

The switch consists merely of a superconducting wire surrounded by a heating coil. It has the interesting property that, though the ratio of the resistances in the open and closed states is infinite, the open-state resistance may be only about an ohm or so. This is because, at low temperatures, even normal metals are very good conductors of electricity. An example of the way that superconducting switches can be used is to be found in Templeton's current-reversing arrangement which is illustrated in Figure 8.3 (Templeton, 1955a). Heaters $A_1$ and $A_2$ are connected to one another, and so are heaters $B_1$ and $B_2$. When power is supplied to the heaters $A_1$ and $A_2$, this drives the superconducting wires that they enclose into the normal state, and the current through the load is passed via the wires enclosed by $B_1$ and $B_2$. It will be seen that the current is reversed when power is supplied to the heaters $B_1$ and $B_2$, instead of $A_1$ and $A_2$. Such a reversing switch is most useful in electrical measurements at liquid helium temperatures.

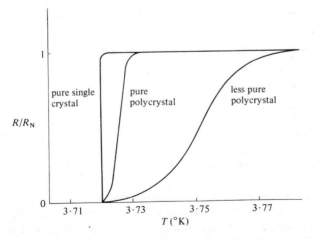

**Figure 8.2.** Plot of resistance $R$ relative to the residual value $R_N$ for the normal state, as a function of temperature, for three samples of tin.

A very simple *helium-level gauge* can be made using a loop of super-conducting wire dipping into the liquid as shown in Figure 8.4. When the level of the liquid helium drops, part of the wire rises to a higher temperature and becomes normal. The measured potential difference across the wire is proportional to the length that is in the normal state.

A *bolometer* has been made successfully by Martin and Bloor (1961), using a sample of superconductor which is held at the transition temperature. The superconductor for such a bolometer should have a very sharp transition to the normal state, as the measured resistance is then sensitive to the very small increase of temperature that results from weak radiation falling on the sample.

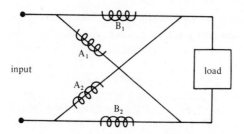

**Figure 8.3.** A superconducting reversing switch.

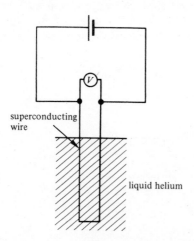

**Figure 8.4.** Liquid helium level gauge.

### 8.1.3 Limited current-carrying capacity of a superconductor

The fact that a superconductor can carry a lossless current only up to a limiting value $I_c$ has been utilised in a device known as the *calotron* (Broom and Rhoderick, 1960). In its simplest form this is a bistable element consisting of a strip of superconductor carrying a current of a certain value $I < I_c$. When an ignition pulse also equal to $I$ is applied, the total current $2I$ becomes greater than $I_c$ and the strip becomes normal. It does not return to the superconducting state when the ignition pulse is removed since the Joule heating effect maintains the temperature above that of the bath. The calotron can, however, be returned to its normal state by applying an extinguishing pulse $-I$, since this removes the Joule effect. In spite of the thermal mechanism that is involved, the switching time is of the order of no more than 1 $\mu$s.

Actually only part of the strip is driven into the normal state by a very short ignition pulse. Further pulses can then be used to make more and more of the sample normal, so that, in effect, the calotron can be used as a multistable element. Broom and Rhoderick have suggested that it could be used as such in digital-to-analogue conversion.

### 8.1.4 The critical-field transition

Most of the electronic applications of type I superconductivity make use of the transition from the superconducting to the normal state in a magnetic field. Although, in principle, this transition can be very steep, it is usually spread out over a rather narrow range of field as illustrated in Figure 8.5.

Suppose now that an alternating magnetic field is superimposed on an applied bias field, so that the overall field is modulated between the limits $B_A$ and $B_B$. The resistance of the sample will then be modulated between $R_A$ and $R_B$. Hence, if a constant voltage is applied to the sample, the current will have an alternating component; the sample is acting as a *modulator*.

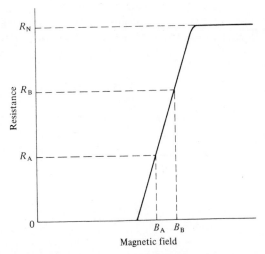

**Figure 8.5.** Resistance plotted against magnetic field in the superconducting normal transition region.

The modulation of the applied field is accomplished by varying the current through a coil. It is easy to see that, when the resistance-field transition is very sharp, a small field modulation can produce a large modulation of the output current. It is, thus, quite easy to make a high-gain superconducting *amplifier*.

Templeton (1955b) has demonstrated that a low-noise *d.c. amplifier* can be made by the combination of a superconducting a.c. amplifier and modulator. The input current is modulated and fed into an audio-frequency transformer that is held, like the other components, in a liquid helium bath. The output from the secondary of the transformer is then applied to the modulating coil of the a.c. amplifier. Further amplification is achieved by conventional means. Such a device is particularly useful in the measurement of very small voltages that originate at liquid helium temperature.

### 8.1.5 Cryotrons and related devices

Although the modulator and amplifiers that have been described are interesting enough, the application of type I superconductivity that has received most attention is the use of the resistance-field transition in computer elements.

The original device was a switching element, the *wire-wound cryotron* (Buck, 1956) that is illustrated in Figure 8.6a. It consists of a straight piece of tantalum wire (the gate) around which is wound a coil of niobium wire (the control). The tantalum can be switched from the superconducting to the normal state by passing a current through the niobium coil, which acts as a solenoid. The gate itself produces a field $\mu_0 I_g/2\pi a$ at its surface, in a direction that is perpendicular to its length, $I_g$ being the gate current and $a$ the radius of the wire. The control provides a field $\mu_0 N I_k/l$ along its axis, where $I_k$ is the control current, $N$ the number of turns, and $l$ the length. The resultant field then has the value of $\mu_0[I_g/2\pi a)^2+(NI_k/l)^2]^{\frac{1}{2}}$.

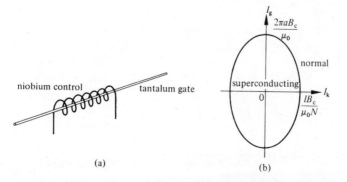

(a)

(b)

**Figure 8.6.** (a) The wire-wound cryotron. (b) The theoretical plot of gate current against control current at the superconducting–normal transition.

Switching occurs when this field is equal to the critical field $B_c$, so the condition for switching is

$$\left(\frac{I_g}{2\pi a}\right)^2 + \left(\frac{NI_k}{l}\right)^2 = \frac{B_c^2}{\mu_0^2} .$$ (8.4)

Figure 8.6b shows the theoretical curve that defines the transition.

It should be noted that the maintenance of the gate in the normal (off) state does not require the expenditure of power if the gate current is zero, since the control has a very high critical field and always remains superconducting.

It is desirable that one should be able to switch a high gate current with a small control current. The ratio of the gate current $I_g$, that switches the cryotron for zero control current, to the control current $I_k$, that leads to switching when the gate current is zero, is called the gain of the device. Wire-wound cryotrons can have gains of up to about 10.

There are two major disadvantages of the wire-wound cryotron. It is not as easy to manufacture and interconnect in large quantities as the thin-film cryotrons (described later), and it is rather slow in operation.

Consider the cryotron flip-flop circuit shown in Figure 8.7. It is supposed that an inductance $L$ is associated with each of the two gates of the two cryotrons and that the gate resistance is equal to $R$ in the normal state. Let the current $I$ be applied when the gate of cryotron 2 is closed (i.e. normal) so that $I_{g_1} = I$ and $I_{g_2} = 0$. The current can be switched to cryotron 2 by applying a current pulse to control 1, the current through control 2 being zero. The voltage drop in each half of the circuit must be the same, so, while gate 1 is normal,

$$I_{g_1} R + L \frac{dI_{g_1}}{dt} = L \frac{dI_{g_2}}{dt} .$$ (8.5)

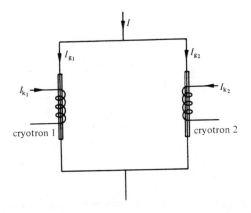

**Figure 8.7.** A flip-flop circuit using a pair of cryotrons.

Integrating up to a time $t$, remembering that $I_{g_1} = I$ and $I_{g_2} = 0$ when $t = 0$, we find

$$\left.\begin{aligned}
I_{g_1} &= I \exp\left(-\frac{Rt}{2L}\right) \\
I_{g_2} &= I\left[1 - \exp\left(-\frac{Rt}{2L}\right)\right]
\end{aligned}\right\}. \tag{8.6}$$

The time constant is thus equal to $2L/R$. It has already been pointed out that the resistance of normal wires tends to be very small at low temperatures. Thus, the inductance must be reduced to an extremely low value too, if fast switching is to be achieved.

It is possible to deposit thin films of niobium and tantalum, but good-quality films of lead and tin are much more easily produced. Thus, a typical *thin-film cryotron* makes use of a lead strip for its control and connectors, and a tin strip for its gate. It is operated at a temperature of about $3 \cdot 65°$K (that is about $0 \cdot 1$ degK below the critical temperature for tin), since the control current that is needed to cause switching is then reasonably small. Figure 8.8 shows the crossed-film arrangement; there is also another type of thin-film cryotron, the in-line cryotron, in which the control and gate are parallel to one another. The in-line cryotron cannot have a gain exceeding unity unless the gate is biased but it can have the advantage of a very low time constant. We consider here the time constant of the crossed-film cryotron.

We expect the length of the normal region of the closed gate to be approximately equal to the width $w_k$ of the control strip. Thus, if $\rho$ is the normal resistivity, the gate resistance is $R = \rho w_k / w_g t_g$, where $t_g$ is the thickness of the gate. The inductance $L$ is associated primarily with the lead connecting strip of length $l$; if the thickness of the insulation between the connectors and the lead ground plane is $t_i$, $L$ is given by $\mu_0 t_i l / w_k$ (it is

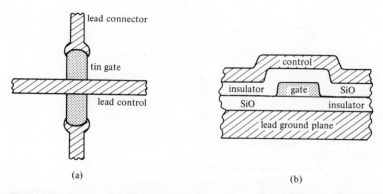

**Figure 8.8.** The crossed-film cryotron: (a) plan; (b) section (the lead ground plane concentrates the magnetic field in the region below the control).

supposed that the connectors and control have the same width). Thus the time constant associated with a single cryotron is

$$\tau = \frac{L}{R} = \frac{\mu_0 t_i t_g w_g l}{\rho w_k^2} .$$ (8.7)

It should be noted that the magnetic field near the surface of a thin conductor of width $w$, carrying a current $I$, is equal to $\mu_0 I/w$. It is, thus, apparent that the gain of the crossed-film cryotron is equal to $w_g/w_k$. Suppose, then, that the gate is made five times as wide as the control and that the latter has a typical width of $10^{-2}$ cm. The length of the element (which we set equal to $l$) may be about 2 cm. For fast switching, the thickness of the films should be as small as possible; in practice films of about 5000 Å might be used. We ignore the effective increase in insulation thickness, associated with the penetration depth in the super-conductor, when estimating $L$. Finally, the resistivity $\rho$ in the normal state is taken to be about $10^{-8}$ Ω m. Substituting the above parameters in Equation (8.7), we find a value for $\tau$ of about $3 \times 10^{-8}$ s. It is clear, then, that very small switching times can be achieved with thin-film cryotrons.

The *Crowe cell*, illustrated in Figure 8.9, is a superconducting memory element (Crowe, 1957). It consists of a lead film in which there is a circular hole bridged by a crossbar. Three insulated lead strips lie parallel to the crossbar; two of these strips are the X and Y drive lines, while the third is the sense line.

A persistent current can be made to flow through the crossbar (returning round the circumference of the hole) in either of two directions. The Crowe cell is said to be in the 'one' state when the current flows in

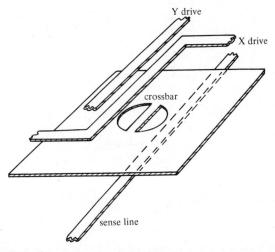

**Figure 8.9.** Exploded view of the superconducting components of the Crowe cell.

one particular direction, and in the 'zero' state when the current flows in the reverse direction. When current is supplied to the drive lines, the crossbar current changes so as to maintain the total current at its original value, so long as the film remains superconducting. In this way the magnetic flux, which comes up one side of the hole and goes down the other side, is kept constant.

Figure 8.10 shows the response of the cell to specified drive pulses. It is supposed that, at time $t = 0$, the cell is in the 'one' state. A current pulse in either direction through either of the drive lines changes the crossbar current $I_z$ momentarily, but $I_z$ returns to its original value at the end of the pulse. Similarly, positive pulses through both of the drive lines at the same time do not change the value of $I_z$ once they have been terminated. However, when the Crowe cell is in state 'one' and negative pulses are applied to both drive lines, the value of $I_z$ exceeds the critical current $I_c$ and the crossbar is driven into the normal state. $I_z$ then decays to the value $I_c$, when the crossbar once again becomes superconducting. At the termination of the current pulses, the crossbar current now flows in a direction that is opposite to the original one; in other words, the cell has been driven into the 'zero' state. In the same way, positive pulses applied to both X and Y lines will drive the cell from the 'zero' to the 'one' state.

The decay of current, while the crossbar is in the normal state, leads to a change of magnetic flux which, in turn, induces an e.m.f. in the sense

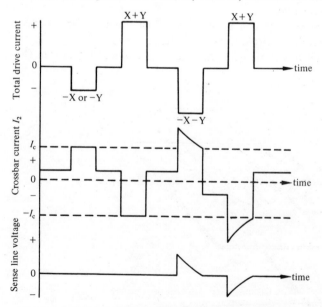

**Figure 8.10.** Diagram showing changes of crossbar current and sense line voltage resulting from drive pulses on the X and Y lines of a Crowe cell.

line. This e.m.f. is negative when the cell is being driven out of the 'zero' state and positive when it is being driven out of the 'one' field. Thus, by applying current pulses to the drive lines one can read out the state of the cell from the signal on the sense line.

### 8.1.6 Thermal switches

The electronic thermal conductivity of a superconductor may be discussed conveniently in terms of the two-fluid theory. According to this theory, some of the electrons in a superconductor have superfluid properties, while the remainder do not, and behave as electrons in a normal metal. The normal electrons do not contribute to the charge transport, since they are short-circuited by the superfluid electrons, but they can, of course, carry heat. On the other hand, the superfluid electrons do not transport thermal energy.

When the temperature lies just below the critical value $T_c$, most of the electrons are still normal in their behaviour and there is little difference between the thermal conductivities in the superconducting and normal states. However, when $T \ll T_c$, nearly all the electrons display superfluid properties and the conduction of heat is due largely to the lattice vibrations. Since, in most normal metals the electronic thermal conductivity is very much larger than the lattice conductivity, it follows that the total thermal conductivity must be much less in the super-conducting state. This rule is sometimes disobeyed for alloys in which the electronic heat conductivity is small even in the normal state but it holds good for pure metals.

The large change in thermal conductivity when a superconductor becomes normal at $T \ll T_c$ is utilised in a low temperature *thermal switch* (Ambler and Hudson, 1955). If a heat source and sink are joined by a piece of superconductor the thermal transfer can be very small, but it can become much greater if the superconductor is driven normal using a solenoid. Ratios of thermal resistance of about 100:1 have been obtained in this way.

## 8.2 Tunnelling effects

### 8.2.1 Single-particle tunnelling

When two metals are separated by a thin layer of insulator it is possible to observe tunnelling of electrons from one metal to the other. If both metals are in the normal state, the current increases more or less linearly with voltage, but, if one or both of the metals are superconducting, the characteristic becomes more interesting.

Consider first tunnelling between a normal metal and a superconductor. Figure 8.11a shows the variations of density of states and electron concentration on either side of the junction at a temperature greater than $0°K$. The current-voltage characteristic at low voltages ($V \ll E_g/e$) is dominated by the fact that there is an energy gap in the density-of-states function for the superconductor. So long as the applied voltage remains much less than $E_g/e$, the tunnelling current is small, because only a small number of electrons on one side of the junction lie opposite empty states on the other side. The number of electrons that can take part in tunnelling rises with voltage but does not become really large until the Fermi level in the normal metal is brought opposite the electron states that lie above or below the energy gap in the superconductor. In other

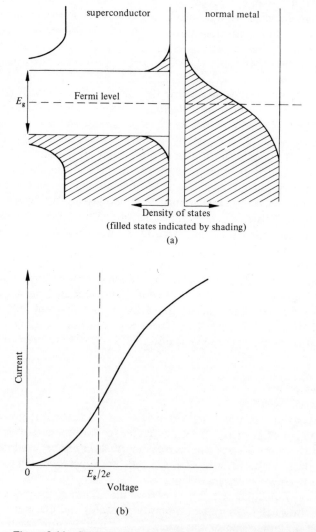

Figure 8.11. Single-particle tunnelling between a superconductor and a normal metal: (a) energy diagram showing density of states and electron concentration on either side of the insulating layer; (b) schematic plot of current against voltage.

words, the current starts to rise very rapidly when the applied voltage rises above the value $E_g/2e$, where $E_g$ is the energy gap. The plot of current against voltage is shown in Figure 8.11b.

Tunnelling between normal metals and superconductors has been used in the determination of energy gaps for the latter materials. However, from the device viewpoint, tunnelling between two different superconductors has more potentiality. Figure 8.12a shows the energy diagram at zero

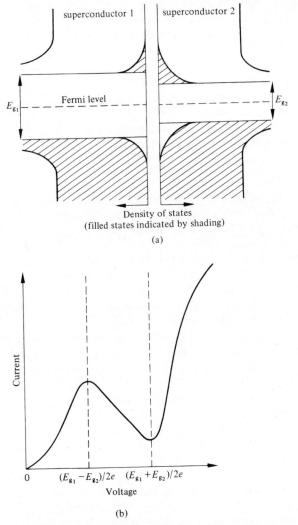

**Figure 8.12.** Single-particle tunnelling between two different superconductors: (a) energy diagram showing density of states and electron concentration in each superconductor; (b) plot of current against voltage.

applied voltage for a junction across an insulating layer between two superconductors designated 1 and 2, where $E_{g_1} > E_{g_2}$. Again there is a very small current when the applied voltage is small; this current rises with increasing voltage until the states just above or below the energy gap, on either side of the junction, lie opposite one another, i.e. when $V = (E_{g_1} - E_{g_2})/2e$. Further rise of voltage reduces the current, since the density of states in the part of the band of superconductor 1 opposite the electrons or holes in superconductor 2 then falls. The current does not rise again until the voltage exceeds $(E_{g_1} + E_{g_2})/2e$, since under this condition the states above the gap on one side of the junction lie opposite states below the gap on the other side. The current-voltage plot has the form shown in Figure 8.12b. It is seen that, between the voltages $(E_{g_1} - E_{g_2})/2e$ and $(E_{g_1} + E_{g_2})/2e$, the differential resistance is negative. One could thus, in principle, make negative-resistance amplifiers from junctions between two superconductors, but it must be borne in mind when considering such applications that the energy gaps are typically of the order of 1 meV. The negative-resistance region of the characteristic occurs at a voltage that is two or three orders of magnitude less than that for a semiconducting tunnel diode.

### 8.2.2 Josephson tunnelling

Particularly interesting effects occur when pairs of superconducting electrons (Cooper pairs) tunnel together from one superconductor to another. The effects were first predicted by Josephson (Josephson, 1962) and are named after him.

The a.c. Josephson effect occurs when a finite voltage $V$ is applied to the junction. As one might expect there is a flow of direct current, but superimposed on this there is a lossless alternating current of frequency $f$ equal to $2Ve/h$ (i.e. 484 Hz $\mu V^{-1}$). When an electron pair tunnels from one superconductor to the other, there is a change of energy $2Ve$ so that a photon of this energy is emitted. The effect has some device potential, since it implies that a voltage measurement can be reduced to a determination of frequency; time and frequency are quantities that can be measured more precisely than any others. There is also the possibility of using the a.c. effect in the detection or generation of microwaves.

The other Josephson effect consists of a direct current which flows even when the voltage across the junction is zero. The voltage drop remains zero up to some limiting current $I_{max}$. It is this d.c. effect which so far seems to offer the greater scope for applications.

A useful configuration is the so-called double junction illustrated in Figure 8.13a. A superconducting loop is divided by thin insulating layers at A and B and Josephson tunnelling can occur at both of these junctions. Now the loop must contain an integral (or zero) number of flux quanta, each having the value $\phi_0$ equal to $h/2e$ ($2 \times 10^{-15}$ Wb). If a gradually increasing magnetic field is applied, a circulating current will flow around the loop, at first opposing the entry of flux. However, when the external field reaches half that value which would exist in the loop if it contained one flux quantum, such a quantum will enter the loop. The circulating current then flows in the opposite direction. The circulating current is zero only when the external field is the same as that which corresponds to an integral number of flux quanta in the loop.

Now consider the largest current $I_x$ from an external circuit that can be carried across the two junctions in parallel. In zero field the current will be equally divided between the two branches, each junction carrying its limiting Josephson current (if we assume that the junctions are well matched). If there is to be a circulating current as well, then the maximum lossless current $I_x$ must fall. $I_x$ will have its largest value when the internal and external magnetic fields are equal and its smallest value just before and just after the entry of a flux quantum. In other words, the maximum Josephson current through the double junction is modulated as the external field changes[1].

Suppose that the area of the loop is $A$ square metres so that the field corresponding to a single flux quantum is $2 \times 10^{-15} A^{-1} T^{-1}$. This then is the periodicity of the modulation of the current $I_x$ as a function of field. By counting the number of oscillations in $I_x$, one can determine changes of magnetic field with a high sensitivity.

A very simple way of constructing a double junction has been described by Clarke (1966). His device, which he calls a 'slug', is illustrated in Figure 8.13b, and is produced merely by dipping a niobium wire into

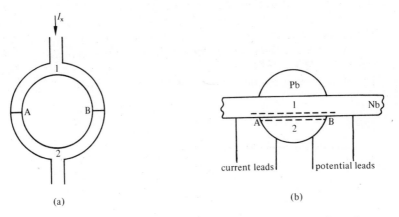

(a)                                         (b)

**Figure 8.13.** The superconducting double junction: (a) ideal configuration; (b) practical realisation (Clarke's 'slug').

[1] In this description the double junction has been idealised, but it behaves qualitatively as predicted. It should be noted that the maximum lossless current through a single Josephson junction is also modulated on applying a magnetic field, but the period of the modulation is much greater than for a double junction.

molten lead, so as to form a lead bead on the wire. This bead is separated from the niobium wire by an oxide film which, it appears, is thinnest at the ends (A and B) so that tunnelling can take place at these positions. The limits of the region that contains the flux quanta are indicated in the figure by the dashed lines; the thickness of this region is of the order of the sum of the penetration depths in niobium and lead.

The most important use of the double junction is in the measurement of very small voltages or currents. It is best employed as a null detector, since the effective area of the loop is sensitive to mechanical strains and changes of temperature. As an example, we show how the double junction is used as a null detector in the determination of the value of a very low electrical resistance.

Figure 8.14 gives the circuit for finding the value of an unknown resistance of $R$ (of the order of $10^{-10}$ $\Omega$) in terms of a standard resistance $S$ (of about $10^{-8}$ $\Omega$). First the field-producing current $I_B$ is set so that the critical current from the niobium to the lead has its mean value, since the sensitivity to change of magnetic field is then greatest. A current $I_R$ is passed through the resistance $R$, there being a corresponding voltage drop $V_R$. The current through the niobium wire of the slug is thus changed and this alters the critical current. The critical current is restored to its original value by backing off the voltage $V_R$ with a current $I_S$ applied across the standard resistance.

One can detect changes in the critical current of a double junction corresponding to about 1% of a modulation period. From this we estimate the smallest change of voltage that might be detectable. Suppose the voltage is applied to a resistance $R$ and the inductance $L$ is so tightly coupled to the double junction that stray inductances can be neglected. The change of current producing an additional flux quantum in the loop is then about $\phi_0/L$, the corresponding voltage change being $R\phi_0/L$. We might therefore hope to detect a change in voltage of about $10^{-2}R\phi_0/L$. Now $L/R$ gives the time constant of the primary circuit which for practical

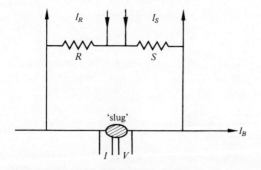

**Figure 8.14.** Circuit for the measurement of very small resistances using a superconducting double junction.

purposes must not exceed, say, 1 s. Substituting the value of the flux quantum, we find a minimum detectable change of e.m.f. equal to about $2 \times 10^{-17}$ V. Actually, Clarke achieved a sensitivity of about $10^{-14}$ V with a time constant of 1 s and this still compares favourably with any other type of detector.

## 8.3 Type II superconductors

The most important application of superconductivity at the present time is undoubtedly in the generation of high magnetic fields without consumption of large amounts of power[2]. This application has come about through the discovery of superconductors that can carry large lossless electric currents in high fields. We shall discuss the physics of these high-field superconductors and then show how they are used in the construction of magnets.

### 8.3.1 High-field superconductivity

The origin of high-field superconductivity lies in the fact that, even in an ideal type I superconductor, the Meissner effect is not quite complete. When the superconductor is placed in a magnetic field, there is flux penetration to a depth $\lambda$. We have already seen that it is the exclusion of flux that makes it energetically more favourable for a superconductor to revert to the normal state in a field greater than $B_c$. However, if the superconductor is in the form of a sheet that is thin compared with the penetration depth $\lambda$, there can be little exclusion of flux, and the super-conducting state is maintained in a field much larger than $B_c$. Thus, one way in which a high-field superconductor can be obtained is by building it up from thin sheets or filaments of type I material that are separated from one another by non-superconducting material. That this approach is feasible has been demonstrated by Bean et al. (1962) who compressed mercury into the interconnected pores of unfired Vycor glass, thus producing a network of mercury filaments each of about 40 Å in diameter. It was found that the mercury-in-glass remained superconducting up to a field of about 0·6 T at 2·16°K, whereas the critical field of bulk mercury is only about 0·03 T at this temperature.

In most high-field superconductors, however, the thin superconducting regions occur naturally. In a pure metal, the surface energy at the boundary between normal and superconducting material is always positive but, according to the Ginzburg-Landau theory, it is possible for this surface energy to become negative for an alloy in a magnetic field. If the surface energy is negative, it becomes favourable for the material to break up into superconducting regions separated by normal regions, this constituting the so-called mixed state.

Since the superconducting regions are thin compared with $\lambda$, the critical

---

[2] A typical water-cooled copper coil of the Bitter type capable of generating a magnetic field of 10 T requires rather more than one megawatt of power.

field for the mixed state rises above the thermodynamic critical field.  If the surface energy is to become negative, it is necessary that the electronic free path in the normal state should be low; this condition is fulfilled for impure metals and alloys.

The effect of alloying is illustrated in Figure 8.15 which shows the results of magnetisation experiments carried out by Livingston (1963). Pure lead is a type I superconductor and shows the full Meissner effect up to the applied field of about $0 \cdot 05$ T at which it becomes normal.  On the other hand the lead–indium alloy retains a substantial (negative) magnetisation up to a field of $0 \cdot 24$ T, although there is appreciable flux penetration at fields in excess of about $0 \cdot 02$ T.  The lead–indium alloy is a type II superconductor and has two critical fields; the lower critical field $B_{c_1}$ is that at which appreciable flux penetration first occurs, while the upper critical field $B_{c_2}$ is that at which the material becomes completely normal [3].

It will be seen that there is a considerable difference between the magnetisation curves for the annealed and cold-worked alloys in Figure 8.15.  The annealed sample shows almost reversible behaviour, while the cold-worked sample displays severe hysteresis, although the upper critical field is the same in both cases.

The practical importance of cold-working becomes apparent when we consider the current-carrying capacity of a type II superconductor. According to Abrikosov, there is a regular arrangement of quantised flux lines in a strain-free type II superconductor, for $B_{c_1} < B < B_{c_2}$, when no current is flowing.  However, when a current flows, the Lorentz interaction pushes the flux lines to one side of the sample.  Such a current can continue to flow only if an e.m.f. is applied; otherwise the flux lines will rearrange themselves in the Abrikosov pattern and quench the current.

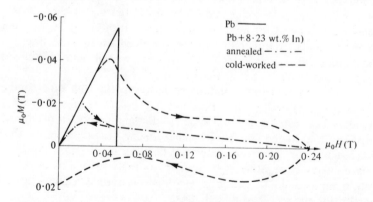

**Figure 8.15.** Magnetisation curves for pure lead and for a lead-indium alloy at $4 \cdot 2°$K.

[3] Actually, if the magnetic field has a component perpendicular to the surface of the material, a thin layer of superconducting phase remains above $B_{c_2}$.

The high current-carrying capacity of useful type II superconductors is due to the presence of defects, notably dislocations. Because of these defects, the energy of a flux line varies with position, so that the lines form flux bundles, lying in energy troughs as shown in Figure 8.16b. Even when a current flows and the energy diagram is tilted as indicated in Figure 8.16c, the flux bundles still remain pinned, though there is clearly a limit to the superimposed energy gradient that can be tolerated.

Actually, at temperatures above $0°K$, there is a small but finite possibility of flux bundles being released from their pinning points by thermal excitation. This means that the current in a type II superconductor is never completely lossless, but for most practical purposes the resistance can be regarded as being zero.

Silsbee's rule, of course, sets an upper limit on the current that can be carried by a type II superconductor, the critical field having the value $B_{c_2}$. However, unless the flux is effectively pinned the observed critical currents can be much smaller than predicted by this rule. In fact the first super-conducting material capable of carrying a high current in a really large magnetic field was not reported until 1961. Then Kunzler $et$ $al.$ (1961) described how they had observed current densities of $10^9$ A m$^{-2}$, for short wires of the compound Nb$_3$Sn in liquid helium, in a magnetic field of $8·8$ T.

It is important to realise that the hysteresis associated with the magnetisation of a type II superconductor of large current-carrying capacity means that alternating current cannot be carried without loss. Thus, the applications of type II superconductors are restricted to those in which direct currents (or very slowly changing currents) are involved.

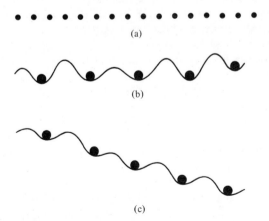

Figure 8.16. Pinning of the flux due to defects in a type II superconductor: (a) schematic representation of the regular distribution of flux lines in a defect-free material; (b) bundles of flux lying at energy minima; (c) flux bundles remain pinned when a current flows.

### 8.3.2 Superconducting solenoids

Superconducting magnets usually take the form of solenoidal coils. The choice of materials for the winding is obviously the first consideration. Besides the intermetallic compound $Nb_3Sn$ that has already been mentioned, the alloys of niobium with zirconium and titanium are also useful superconducting magnet materials. $Nb_3Sn$ has an upper critical field in excess of 20 T and can carry very large currents at lower fields, but is very brittle. Niobium-titanium and niobium-zirconium are often preferred, in spite of upper critical fields of only $14 \cdot 5$ T and 13 T respectively, on account of their reasonable ductility.

It seems likely that upper critical fields of the order of 40 T will not be exceeded in the foreseeable future. Clogston (1962) has pointed out that the fact that a normal metal is paramagnetic is an inducement for a superconductor to enter the normal state in a high magnetic field. Because their susceptibilities are so low, the change of energy of paramagnetic materials is negligible at moderate field strengths. However, in high fields the negative energy change due to the positive magnetisation in the normal state must be taken into account. This means that, as the magnetic field strength is increased, the normal-state energy will eventually fall below that of the superconducting state, quite apart from the effect of flux exclusion which raises the superconducting energy. The paramagnetic susceptibility of a normal metal is $\chi = 2\mu_0 N\beta^2$, where the g factor has been taken to be equal to 2; $N$ is the density of states per unit energy and $\beta$ is the Bohr magneton. Thus, the change in energy of magnetisation per unit volume in a field $B$ is $-\chi B^2/2\mu_0 = -N\beta^2 B^2$. According to the Bardeen–Cooper–Schrieffer theory, the difference between the free energies per unit volume in the superconducting and normal states, in zero field at $0°K$ (where the difference is greatest), is given by $NE_g^2(0)/8$, $E_g(0)$ being the energy gap at $0°K$. Moreover, $E_g(0)$ is equal to $3 \cdot 5kT_c$, so the limiting value of the critical field is given by $\beta^2 B_{c_2}^2 = (3 \cdot 5kT_c)^2/8$. Substituting the value of the Bohr magneton, we find that $B_{c_2}$ has a limit of about $2T_c$ teslas. It appears that no superconductor will be found with $T_c$ much greater than about $20°K$ ($Nb_3Sn$ has an initial temperature of about $18°K$). Thus, we do not expect $B_{c_2}$ ever to be much greater than about 40 T.

The winding of a niobium-zirconium or niobium-titanium solenoid is fairly conventional, but it is often found that the design must be such that the working current is appreciably less than the critical current indicated by tests on short samples. It must be borne in mind that, as the current is increased, the flux bundles which are weakly held at pinning points become freed and the phenomenon of flux creep occurs. Flux creep implies that heat is generated and, if it is not removed, the material can become normal. Thus, the superconducting wires are often covered with copper which assists in the dissipation of heat (copper being a much better conductor of heat than any high-field superconductor) and also provides an alternative low-resistance electrical path, should a small section of the winding become normal. Furthermore, the power supply is designed so that the current through the solenoid can be changed in a slow and regular manner, otherwise catastrophic 'flux jumping' will occur.

The construction of $Nb_3Sn$ solenoids is rather more difficult, although flexible strips of wire supporting thin layers of the superconducting compound are now available. One of the early $Nb_3Sn$ magnets was made from a composite wire of niobium and tin, which was first wound into solenoidal form and then heat-treated so that the elements reacted with one another to form the compound.

A feature of many superconducting solenoid systems is the provision of a so-called persistent switch. This is a length of superconducting wire that bridges the ends of the solenoid and which can be driven normal by the application of heat. Whenever any adjustment of the current through the magnet is to be made, the switch is driven normal. When the switch becomes superconducting once more, it isolates the solenoid from the external power supply, which can then be disconnected without affecting the magnetic field. An advantage of the persistent switch is that it reduces the consumption of liquid helium since, except when the current through the solenoid is being altered, there is no Joule heating in the wires that enter the cryostat.

### 8.3.3 Inductive methods for the magnetisation of superconductors
The difficulty of making superconducting magnets from brittle materials like $Nb_3Sn$ would be eased if the coil could be machined from the bulk material, but this implies a very small number of turns and a very large current. In the limit, of course, the magnet would consist of a single turn, that is a ring or tube.

The magnetisation of a superconducting tube is simple enough if a high-field magnet is available. For example, the field from such a magnet can be applied to the tube when the latter is in the normal state, above its critical temperature. Then, when the temperature is reduced below $T_c$, the flux within the tube becomes trapped and the external field can be removed. Alternatively the tube can be magnetised by applying a rather higher field while it is in the superconducting state. As the external field rises, a current circulates through the wall of the tube so as to oppose the entry of flux, but, once the critical current is reached, flux begins to enter the bore (it must be remembered that there is only partial flux exclusion for a type II superconductor). When the external field is reduced to zero, circulating currents flow in the opposite sense round the tube so as to oppose the exit of flux from the bore. The magnitude of the field that can be trapped in the tube depends on the thickness of the wall; for example, a field of 5 T can be contained in a sintered $Nb_3Sn$ tube with a wall thickness of a few millimetres.

There are certain techniques available for increasing the intensity of the field trapped in a superconducting cylinder. One of the simplest is shown in Figure 8.17. In this device, described by Swartz and Rosner (1962), the flux, which is trapped in the two interconnected holes, is driven into the smaller hole by forcing the superconducting piston into the larger hole. At first sight it might appear that the field would be increased by the ratio of the total cross-section area of the holes to the cross-section area of the smaller hole. However, the performance is rather inferior to this, since account should really be taken of the fact that the flux in the walls also rises during the compression process.

There is an intermediate stage between the thin-wire solenoid with its conventional power supply and the single-turn flux compressor. This is the solenoid with relatively few windings of large cross-section area, which needs a large current to produce a high magnetic field. The introduction of a large current, through what must be leads of normal metal, into a helium cryostat presents considerable problems. It would be much more convenient if the wires entering the helium enclosure carried a much smaller current, which might then be increased through the solenoid itself by an inductive method. This is the idea behind Laquer's flux pump (Laquer, 1963) that is illustrated in Figure 8.18. All the wires shown in diagram other than the leads to the primary of the transformer are superconducting. $S_1$ and $S_2$ are persistent switches, as described previously, that can be driven normal by heating coils. At first $S_1$ is closed and $S_2$ is open. The current in the $n$-turn primary of the transformer is raised from zero to a value $I$. Then, when switch $S_2$ is closed and the primary current is reduced to zero, a current equal to $nI$ is

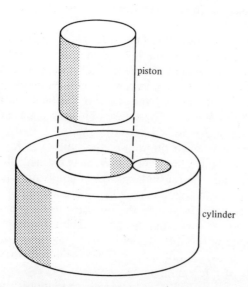

piston

cylinder

**Figure 8.17.** The superconducting flux compressor.

induced in the single-turn secondary. Switch $S_1$ is successively opened and closed, whereupon a persistent current equal to $nIL_2/(L_1+L_2)$ flows through the solenoid, where $L_1$ is the inductance of the solenoid and $L_2$ the inductance of the auxiliary loop. Repeated operation of this cycle of events leads to a current $nI$ through the solenoid.

A very ingenious way of inducing current to flow in a superconducting coil without external current leads has been described by Volger and Admiraal (1962). They made use of the device, a superconducting unipolar dynamo, that is shown in Figure 8.19. A normal 'hole' is formed in the superconducting disc by cooling it from above its critical temperature with a permanent magnet held in the position indicated. Since the disc is made from a type I superconductor, the flux through the

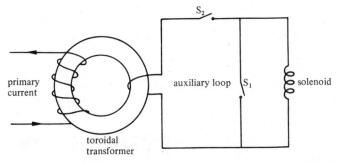

**Figure 8.18.** An electrical flux pump for use with high-current superconducting solenoids.

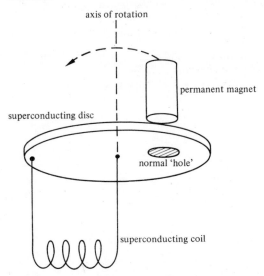

**Figure 8.19.** Induction of current in a superconducting solenoid using a unipolar dynamo.

hole must remain constant. The magnet is rotated about the axis of the disc, causing the normal hole to rotate with it. Each rotation of the magnet leads to an increase in the flux through the superconducting loop composed of the coil, its leads, and part of the disc. The flux contained in the loop, and consequently the current in the coil, can, in principle, be raised to any required level by continual rotation of the magnet (provided that no part of the circuit is driven into the normal state by the process). Volger and Admiraal reported that they were able to raise the current through a niobium-zirconium coil from zero to 50 A, using lead as the type I superconductor for the disc.

### References

Ambler, E., and Hudson, R., 1955, *Rep. Progr. Phys.*, **18**, 251.
Bardeen, J., Cooper, L. N., and Schrieffer, J. R., 1957, *Phys. Rev.*, **108**, 1175.
Bean, C. P., Doyle, M. V., and Pincus, A. G., 1962, *Phys. Rev. Letters*, **9**, 93.
Broom, R. F., and Rhoderick, E. H., 1960, *Solid-State Electron.*, **1**, 314.
Buck, D. A., 1956, *Proc. Inst. Radio Engrs.*, **44**, 482.
Clarke, J., 1966, *Phil. Mag.*, **13**, 115.
Clogston, A. M., 1962, *Phys. Rev. Letters*, **9**, 266.
Crowe, J. W., 1957, *IBM J. Res. Develop.*, **1**, 295.
Fairbank, W. M., Pierce, J. M., and Wilson, P. B., 1963, *Proc. 8th Intern. Conf. Low Temp. Phys.* (Butterworths, London), p.324.
Josephson, B. D., 1962, *Phys. Letters*, **1**, 251.
Kunzler, J. E., Buehler, E., Hsu, F. S. L., and Wernick, J. E., 1961, *Phys. Rev. Letters*, **6**, 89.
Laquer, H. L., 1963, *Cryogenics*, **3**, 27.
Livingston, J. D., 1963, *Phys. Rev.*, **129**, 1943.
Martin, D. M., and Bloor, D., 1961, *Cryogenics*, **1**, 159.
Meissner, W., and Ochsenfeld, R., 1933, *Naturwissenschaften*, **21**, 787.
Pippard, A. B., and Pullan, G. T., 1952, *Proc. Cambridge Phil. Soc.*, **48**, 188.
Swartz, P. S., and Rosner, C. H., 1962, *J. Appl. Phys.*, **33**, 2292.
Templeton, I. M., 1955a, *J. Sci. Instr.*, **32**, 172.
Templeton, I. M., 1955b, *J. Sci. Instr.*, **32**, 314.
Thomas, E. J., 1965, *Brit. Commun. Electron.*, **12**, 56.
Volger, J., and Admiraal, P. S., 1962, *Phys. Letters*, **2**, 257.

# List of symbols

A    (subscript) acceptor
$A$    area, constant in Hall equation
B    (subscript) base
$B$    magnetic induction
$B_c$    critical field of superconductor
C    (subscript) collector
$C$    capacitance
$c$    velocity of light
D    (subscript) donor
$D$    diffusion coefficient, detectivity, electric displacement
$d$    depth, piezoelectric cross-coupling term
E    (subscript) emitter
$E$    energy of carriers
$E_g$    energy gap
$\mathscr{E}$    electric field
$\hat{\mathscr{E}}$    peak field
$\mathscr{E}_B$    breakdown field
$\mathscr{E}_c$    ferroelectric coercivity
$e$    electronic charge
$f$    frequency, occupation number
$f_L$    Larmor frequency
$G$    free energy
$g_m$    mutual conductance
$H$    magnetic field
$H_c$    coercivity
$H_k$    anisotropy field
$h$    Planck's constant ($2\pi\hbar$)
$I$    current
i    (subscript) intrinsic
$j$    current density
$K$    piezoelectric coupling constant
$k$    Boltzmann's constant, magnetic anisotropy constant, piezoelectric coupling coefficient
$\mathbf{k}$    wave vector
L    (subscript) lattice, load
$L$    inductance
$l$    length
$M$    magnetic moment per unit volume
$m$    mass of free electron, magnetic moment
$m^*$    effective mass
N    (subscript) normal
$N$    impurity concentration
$N_v$    number of valleys
n    (subscript) n-type

| | |
|---|---|
| $n$ | electron concentration, refractive index |
| $P$ | photon flux, volume polarisation |
| p | (subscript) p-type |
| $p$ | hole concentration |
| $Q$ | charge |
| $Q_s$ | excess charge |
| q | phonon wave vector |
| $R$ | resistance |
| $R_H$ | Hall coefficient |
| S | (subscript) superconducting |
| $S$ | mechanical strain |
| $s$ | elastic constant |
| $T$ | absolute temperature, mechanical stress |
| $T_c$ | critical temperature, Curie temperature |
| $t$ | time, thickness |
| $U$ | energy |
| $V$ | voltage |
| $V_B$ | breakdown voltage |
| $V_c$ | contact potential difference |
| $V_{app}$ | applied voltage |
| $v_s$ | velocity of sound |
| $W$ | power |
| $w$ | width |
| $Z$ | impedance |
| $z$ | thermoelectric figure of merit |
| $\alpha$ | Seebeck coefficient, attenuation constant, fraction of emitter current reaching collector |
| $\beta$ | Bohr magneton, thermoelectric quality factor, current gain of transistor |
| $\gamma$ | magnetogyric ratio |
| $\delta$ | loss angle |
| $\epsilon$ | permittivity, emissivity |
| $\epsilon_0$ | permittivity of free space |
| $\epsilon_r$ | relative permittivity |
| $\epsilon_{re}$ | electronic contribution to relative permittivity |
| $\eta$ | reduced Fermi energy, quantum efficiency |
| $\theta_D$ | Debye temperature |
| $\kappa$ | thermal conductivity |
| $\lambda$ | wavelength, penetration depth, scattering parameter |
| $\mu$ | mobility, permeability |
| $\mu_0$ | permeability of free space |
| $\mu_H$ | Hall mobility |
| $\pi$ | Peltier coefficient |
| $\rho$ | resistivity, charge density |

| | |
|---|---|
| $\sigma$ | electrical conductivity, Stefan's constant |
| $\tau$ | recombination time, switching time, time constant, torque |
| $\tau_0$ | response time |
| $\phi_0$ | flux quantum |
| $\chi$ | susceptibility |
| $\chi_P$ | Polder tensor |
| $\omega$ | angular frequency |

# Index